深谷光軌 [編著]
倉橋潤吉+山口詳二朗 [編集協力]

深谷光軌作品集
「外空間」を創る

Design of Koki FUKAYA
INNER-LANDSCAPE DESIGN
written & edited by Koki FUKAYA
editorial cooperation by Junkichi Kurahashi & Shojiro Yamaguchi

マルモ出版
Marumo Publishing

目次　　　　　　　　　　　　　　　ギリシャの旅から......6

第1章
外空間ワークス
......13

プロローグ......14

京王プラザホテル［北館］4号街路　外空間　1971......20

日野自動車工業本社「日野台の杜」外空間　1972......36

「日野台の杜」......43

「日野の湧水」......44

ホテルグランドパレス　外空間　1972......54

小西酒造東京支店「坪庭」外空間　1974......60

東京銀行青山寮　外空間　1975......64

能登「サンかがや」外空間　1976......68

東銀栗田ビル（現一ツ橋SIビル）外空間　1978......74

NTT広島仁保ビル　外空間　1980......86

工学院大学八王子校舎5〜11号館　外空間　1986......106

工学院大学八王子校舎3号館　外空間　1987......118

スコープ本社GLASS HOUSE　外空間　1991......128

三番町KSビル　外空間　1991......134

第2章
創作ノート
......151

立っている精神　1979......152

厚別商業センター　外空間　1977......154

朝日生命成人病研究所　1968......156

イトーヨーカドーグループ
多摩研修センター　外空間　1986......157

レストランあかべこ　外空間　1974......158

松方邸　1972......159

小寺邸　1976......160

水野邸　1987......161

スケッチ......162

エスキース......164

付録
ディテール図面編
......165

データ......188

エピローグ......190

プロフィール......191

Contents

From a trip to Greece 6

Chapter.1

Inner-Landscape Works
......13

Prologue 14

Keio Plaza Hotel (North Wing.), Street No.4 *'Gai-Kukan'* 1971......20

Hino Motors, Main Office, 'Forest of Hinodai' *'Gai-Kukan'* 1972......36

'Forest of Hinodai' 43

'Spring Water of Hino' 44

Hotel Grand Palace *'Gai-Kukan'* 1972......54

Konishi Brewing Company Tokyo Branch Office 'Tsuboniwa' *'Gai-Kukan'* 1974......60

The Bank of Tokyo Aoyama Retreat *'Gai-Kukan'* 1975......64

Sun Kagaya in Noto *'Gai-Kukan'* 1976......68

Togin Kurita Bldg. (the present is Hitotsubashi-SI-Bldg.) *'Gai-Kukan'* 1978......74

NTT Hiroshima Niho Bldg. *'Gai-Kukan'* 1980......86

Kogakuin University Hachioji Campus, Bldgs. No.5 to 11 *'Gai-Kukan'* 1986......106

Kogakuin University Hachioji Campus, Bldg. No.3 *'Gai-Kukan'* 1987......118

SCOPE Main Office GLASS HOUSE *'Gai-Kukan'* 1991......128

Sanban-cho KS Bldg. *'Gai-Kukan'* 1991......134

Chapter.2

Notes on Creation
......151

Spirit of Standing Still 1979......152

Atsubetsu Shogyo Centre *'Gai-Kukan'* 1977......154

The Institute for Adult Diseases, Asahi Life Foundation 1968......156

Ito-Yokado Group Training Center in Tama *'Gai-Kukan'* 1986......157

Akabeko Restaurant *'Gai-Kukan'* 1974......158

Matsukata Garden 1972......159

Kodera Garden 1976......160

Mizuno Garden 1987......161

Sketch 162

Esquisse 164

Appendix

Details
......165

Data......188

Epilogue......190

Profile......191

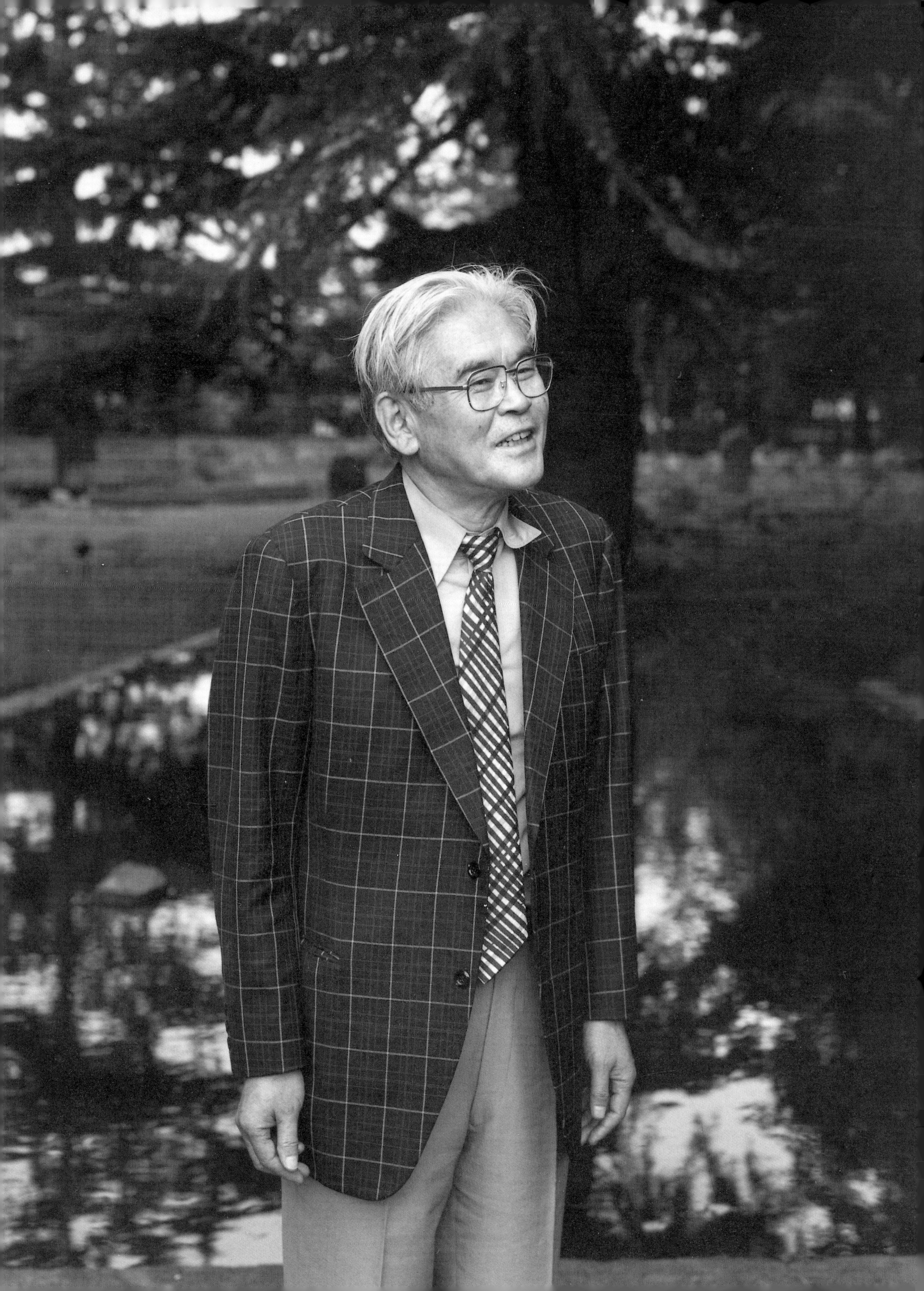

ギリシャの旅から

　私はもともと寺の住職になる予定だった。幼少の時に心臓弁膜症を患って、それで寺を継ぐことを断念し学校出てからある職業についた。けれど自分のやっている仕事に情熱が持てない。それで仕事を辞めた後、弁膜症の手術をして1年間の入院生活を送った。1880年〜90年そこで出合った本がブルーノ・タウトの「日本文化私観」だった。一部を抜粋して要約すると「日本の神社は森の中にある。森に囲まれて神社がある。墓も手入れの行きとどいた、磨きたてられた石ではない。むしろ自然に任せて自然の中に溶け込む自然の中に消滅し去っていくもののように見える。殊に田舎の墓地は、特定の範囲に限られることはなく、樹の下とか、その他任意の一隅に散在し、全くの自然の風景の中に姿を消してしまい、いわば死者の霊も身体も大自然の中に吸収融合されてしまうようになっている。日本の神は自然を超出している神、樹木に囲まれ、自然と共に在る神である。日本人は自らの一生を大地から生まれ、やがて大地に帰り、自然の中へ消滅するものとして、少なくとも近代以前は理解していた。」とても感動し、この一文は私のこころの内面を強く動かした。

　ブルーノ・タウトの本が出会いとなり、自分の独自の分野を開拓していこうと考えた。

　最初はつくっていくものが造園と称する庭としか理解されなかった。どうやって抜け出すか日々葛藤の中で苦しみ悩んだ。

　自分はこれがつくりたいという事を口にしたくない。いろいろな葛藤の末、1970年にギリシャの旅に出た。アテネのアクアポリスのパルテノン神殿の創造の偉大さ、コリント、ミケーネなどの遺跡、その岩や石にへばりつくように咲いている草花の情景が描く破壊された美は、終生忘れ得ることのできない大きな感動であった。もともと自然の中にあった素材が人工的に加工されて、偉大なものが出来上がり、それが風雪に耐え戦火と共に元の自然に帰っていく。その偉大な遺跡が魅せる破壊の美と、可憐に咲くヒナゲシの素朴な生が織り成す悲哀の美は、そこに身を落ち着ける私の心を静めた。ギリシャの旅での体験はその後の私の「外空間」創造の仕事に確信を与える事となった。その確信が京王プラザホテル、日野自動車工場本社、その後手掛けたいくつかの外空間の仕事につながっていった。

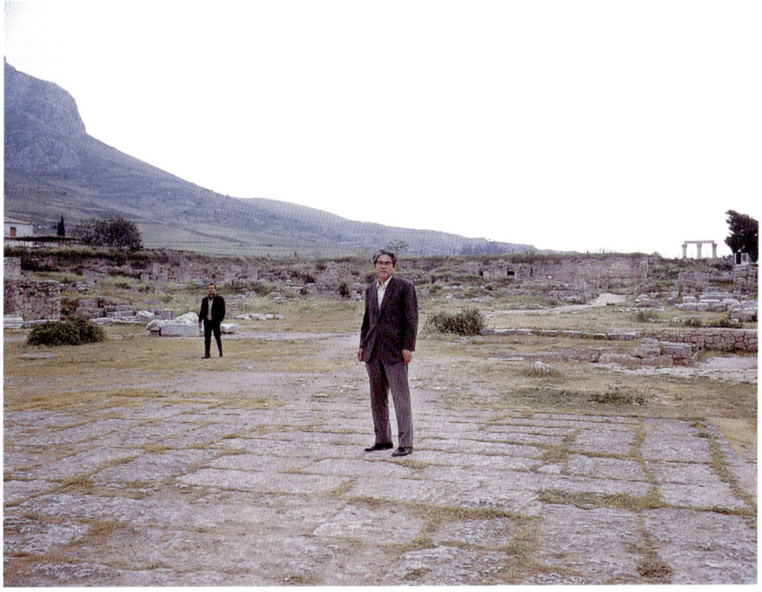

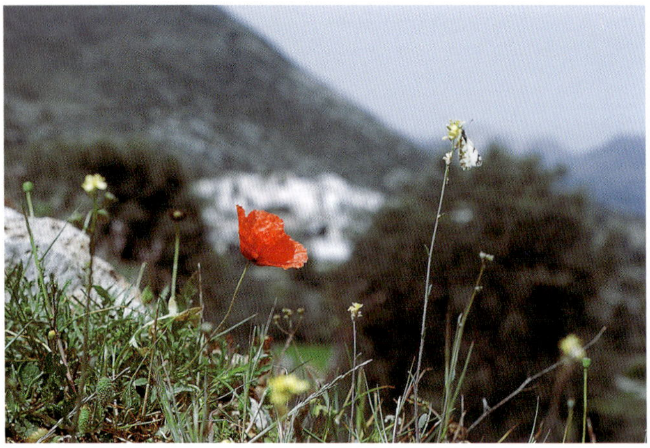

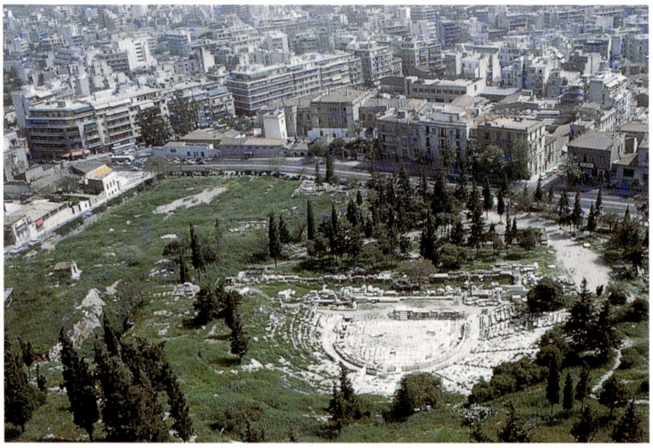

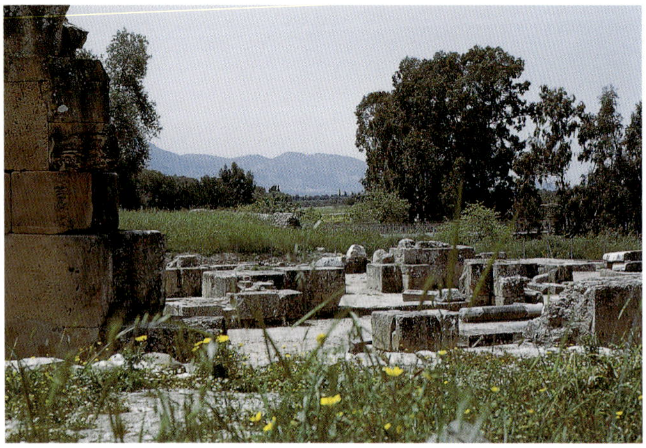

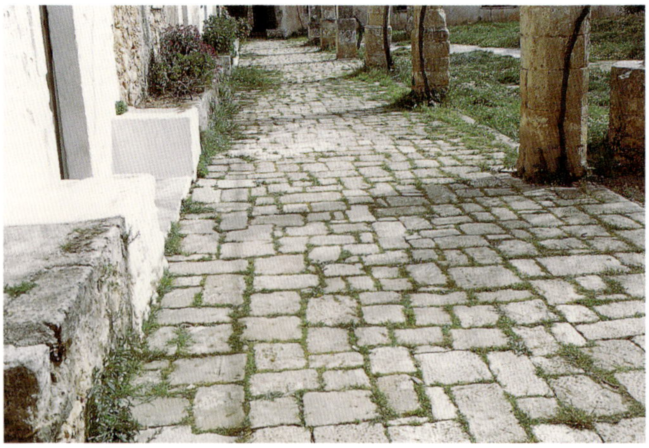

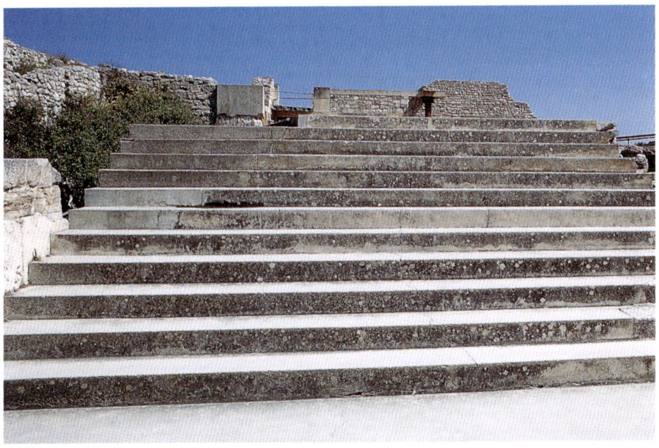

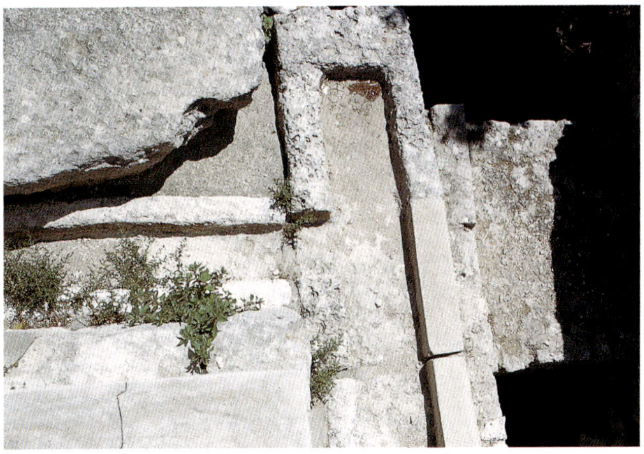

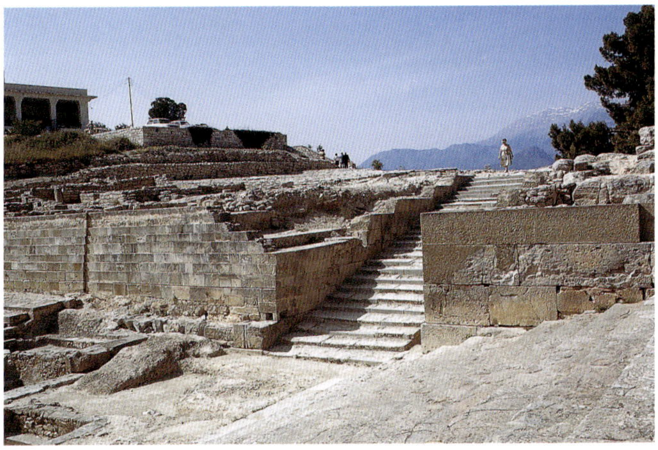

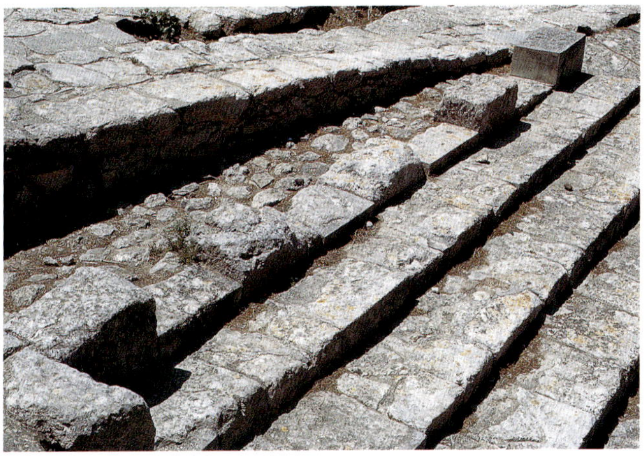

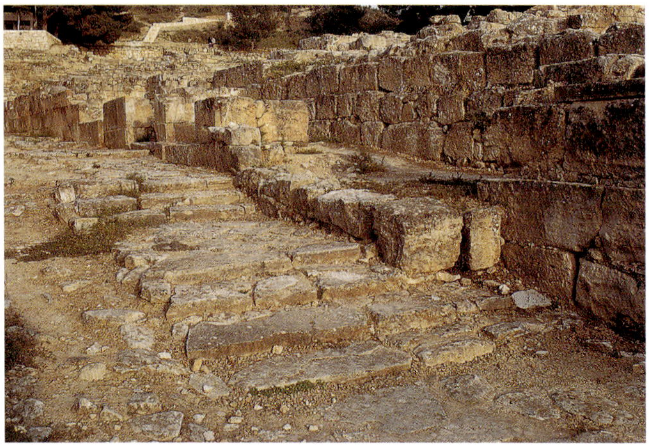
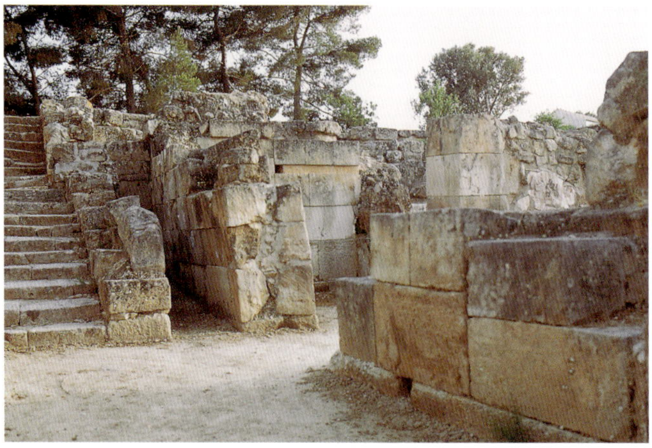
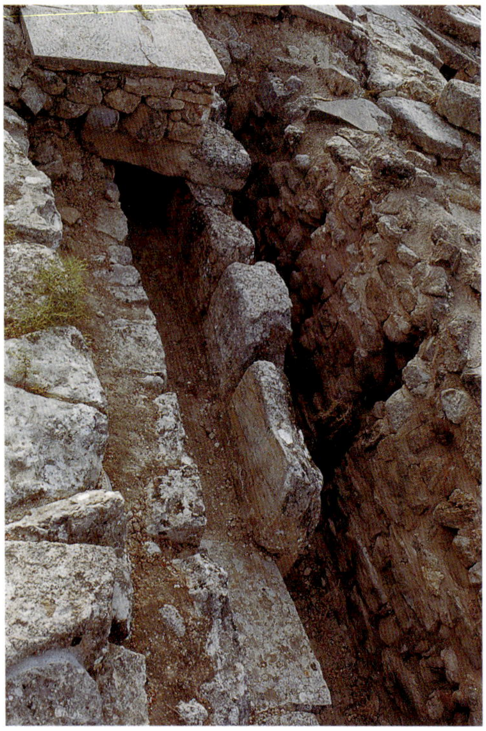
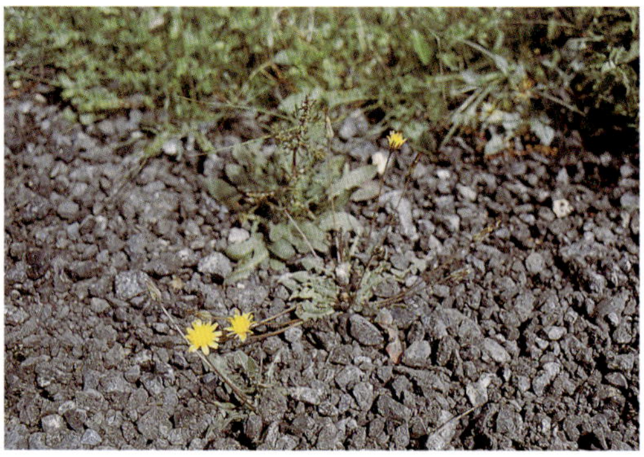

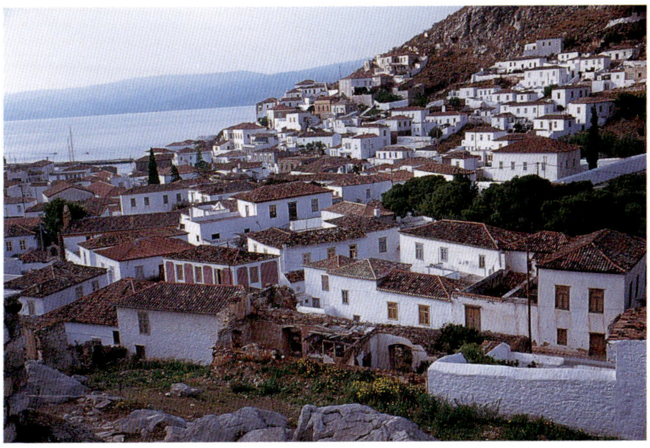

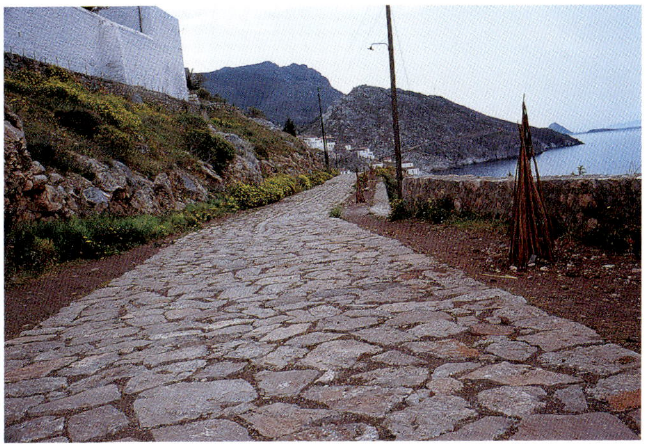

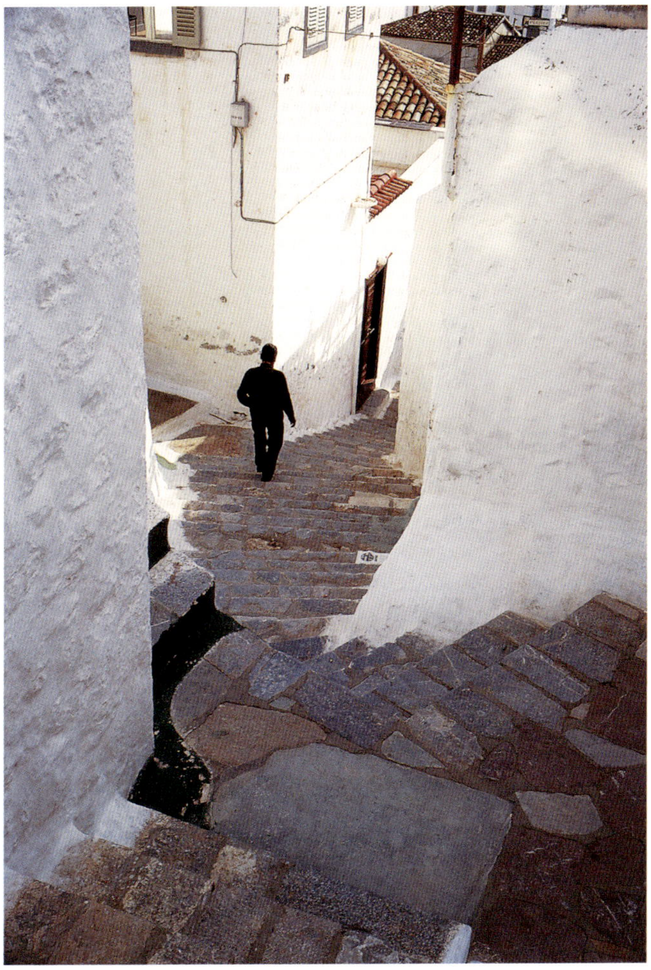

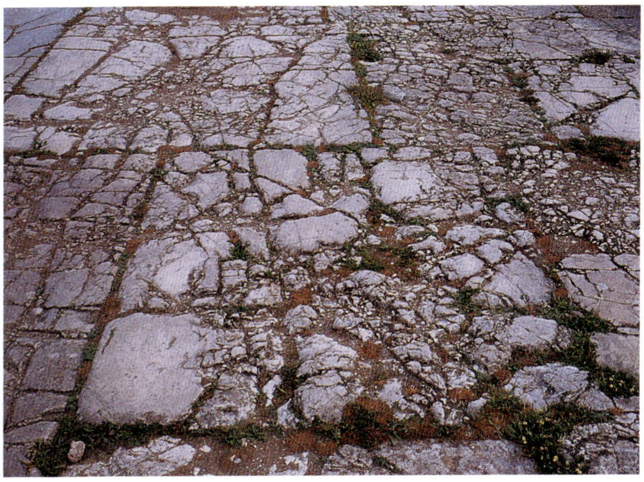

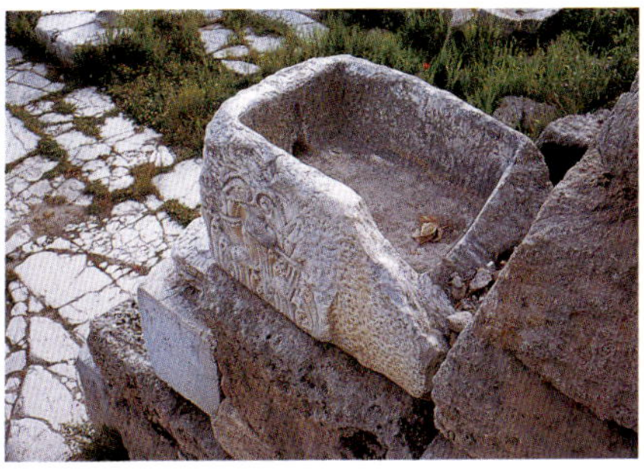

From a trip to Greece

I was originally going to be a Buddhist monk.

Having suffered from valvular heart disease as child, so I gave up the plan to succeed my father's temple and got a job after school. However, I could not get enthusiastic that work so I quit my job and spent a year in hospital after operation for the valvular heart disease. Those days in 1980-1990, I came across the book that was *'A Personal Observation on Japanese Culture - Nippon, Japan seen through European Eyes'* by Bruno Taut (1880-1934) excerpted and summarized as follows; 'Japanese shrines are built in woodlands, and the graveyards in the woodlands are not made up of well-kept or polished tombstones; as if they were just left it to nature, being merge into nature gradually, and disappear last. Especially the graveyards in the countryside are not limited to any designated area, but it can be found under some trees or in other unspecified spaces. It was totally lost from viewing in nature, in a sense as if the body and soul was in harmony with great natures. Japanese god be transcend to nature, among trees and also coexist with nature. Japanese recognized the death as which be born from the earth and go back to the earth, and disappear into nature last, before present day.' I was deeply impressed and this sentence has influenced my inner heart greatly.
Reading Bruno Taut's book changed my outlook on life, and I decided to set out to break ground in my own field.

At the beginning, my creations were understood as only 'Zouen' referred gardens. So I was trying to solved my mind conflict and to alleviate caused daily agony.

I would not prefer to vocalize what would like to create. After long and careful consideration, I had deep impression from those breathtaking places that the magnificent creations of the Parthenon on Acropolis in Athens, ruins of Corinth and Mycenae, the beautiful view of wildflowers were Blossommed on those ruined rocks and stones, these are unforgettable. At first natural materials in which naturally existed, next were processed into something masterpiece by humans, and after time endured weather-beaten and damaged by wars. At last returning to nature for long time past. So to speak the weathered beauty of the great ruins and the tragic beauty of the dainty corn poppies soothed my mind on sight. My experiences from the trip to Greece gave me strong confidence in my creative work of '*Gai-Kukan*' later. That confidence led to some of my other exterior works such as the Keio Plaza Hotel and Hino Motors, Main Office.

第1章　Chapter.1
Inner-Landscape Works
外空間ワークス

プロローグ

　20 数年前になろうか。都市の外部空間を表現空間に選んで、自身に規律を課し、素材と造形手法を厳しく限定して、一定の空間体験に近づこうとした。

　こうして創り出した空間を外空間と名付けた。外空間の名付親は故人の浜口隆一氏である。

　以来 20 数年間、私なりに都市の外空間の創作を重ねてきたがその 20 数年の時の流れの中で、次の 5 つの空間が、当初の目的とした「空間が道と一体化された空間」、即ち道がふくれ、まわりの建物のファサードを効果的な背景として人びとを快適に抱き込むような空間に近付き得たと思われる空間である。

　1971 年　京王プラザホテル 4 号街路　外空間

　1978 年　東銀栗田ビル　外空間

　1980 年　NTT 広島仁保ビル　外空間

　1991 年　スコープ本社 GLASS HOUSE　外空間

　1991 年　三番町 KS ビル　外空間

　以上のそれぞれの空間は、自然を愛する日本人の昔ながらの心が形を変え素材を変えて、現代の冷たい空間と現代人の乾いた気持に割り込めるだけの強さを持っていると、自負している。

　目指したのは、迫力ある外空間を創り出すことであった。そのために空間構成は、「自然な自然」と「意図された逞しい自然」の組み合ったものがよかろうとした。

　「自然な自然」とは、穏やかな流水と転石、風に風を見せる落葉樹に代表される植物群。

　「意図された逞しい自然」とは、幾何学や方向性を感じさせる敷石や石組、激しい流水に逆らって突き出た石、などである。

　以上の 5 つの空間のほか、創り出された空間が敷地の内部に取り込まれているものがある。

　その代表的な空間は、

　1971 年　京王プラザホテル 7F 屋上庭園

　1972 年　日野自動車工業本社「日野台の杜」

　1972 年　ホテルグランドパレス　外空間

　1973 年　王子スポーツガーデン　外空間（現在消滅）

　1974 年　レストラン「あかべこ」横浜店　外空間（現在消滅）

　1974 年　小西酒造東京支店　外空間（現在消滅）

　1975 年　小西酒造本店貴賓室　外空間

　1975 年　東京銀行青山寮（クラブハウス）　外空間

　1976 年　和倉温泉「かがや」別館「サンかがや」　外空間

　1977 年　札幌副都心公社「厚別商業センター」　外空間（一部消滅）

外空間の一部

　1986 年　イトーヨーカドーグループ多摩研修センター　外空間

　1986 年　工学院大学八王子校舎 5 号館〜 11 号館群　外空間

　1987 年　工学院大学八王子校舎 3 号館　外空間

住　宅

　1972 年　松方邸　にわ

1976 年　小寺邸　にわ
　　1987 年　水野邸　にわ
　　1977 年　イラン国ハダザデ邸ゲストハウス　壁泉
　以上のそれぞれの空間の創成への過程、ディテールなどは、前の 5 つの空間を創り出す過程に多くの手掛かりを生んでいる。
「意図された逞しい自然」の素材としての石
　鉄・ガラス・コンクリートに代表される、硬質で乾いた建築群に囲まれた冷たい現代都市空間に、思考する迫力ある空間を創り込ませる為に、外空間を構成あるいは展開する「意図された逞しい自然」としての素材は、建築群に使われた硬質で乾いた素材のそれぞれと同等の強さを持つ異質性が必要とされた。
　当初 1965 ～ 80 年は
（1）階段・舗床
　　　階段、段　石……花崗岩　稲田石（硬質）
　　　板　石……花崗岩
　　　板　石……花崗岩　塩山石（硬質）
　　　小舗石……花崗岩　福島石（硬質）
　　　小舗石……花崗岩　中国産
（2）石積（堅固さを要する石積）
　　　切　石……安山岩　新小松石（中硬石）
　　　　　　　　安山岩　白河石黒目（中硬石）
　　　間知石……安山岩　新小松石（中硬石）
　　　　　　　　安山岩　赤城真石（中硬石）
　　石積（化粧石積）
　　　切　石……角閃石・石英・両輝安山岩
　　　　　　　芦野石（軟　石）
（3）土留・化粧ボーダー切石……新小松石
　　　　　　　　　　　　　　　　芦野石
　　　転石　雑石……新小松石
　　　　　　　　　芦野石
　　　石組　雑石……新小松石
　　　　　　　　　芦野石
（4）単体……新小松石
（5）石の構築との対比を必要とされる箇所にコンクリート打放し、ツツキ仕上
　　1981 ～ 91 年
（1）階段・舗床……花崗岩　稲田石
　　　コンクリート打放し、ツツキ仕上壁笠石……花崗岩　稲田石
（2）石積　切　石……花崗岩　ラステンバーグ

間知石……安山岩　新小松石
（３）化粧ボーダー切石……花崗岩　ラステンバーグ
　　　転石　切　石……花崗岩　ラステンバーグ
　　　土留ボーダー切石……長尺３～５ｍ　芦野石
（４）単体……安山岩　新小松岩
（５）その他　コンクリート打放し、ツツキ仕上

　1981年以降、石種が若干変わったのは、
　　建物の外壁の被覆素材が、時の経過の試練へ向けて持続性が要求されて、石が主流となりつつあることに対応して、それと同等の強さを持った異質性の目的を満たす石種に変えたことによる。
　　それまでの新小松石の素材、加工その他のコストの高騰化及び芦野石に、私の使用する素材として欠陥が発見されたためである。
　　但し、芦野石の欠陥は私の制作の上でのことと、私の考える時の経過に対する持続性の問題であることを断っておく。

制作過程のこと
　各時代の空間の図面と写真をよく見て戴くと理解が生まれると考えるが、A 1971～80年に制作された外空間と、B 1981～91年のそれとでは、大分趣が変わっていることに気付くであろう。その趣のちがいは、
A．1971～80年
①多くの場合、個々の制作現場に、制作作業のための石加工の下小屋を設ける用地の確保が比較的容易で、切削機を含む工作機械の使用が、天候その他に関係なく可能であったこと。
②切石積、土留、化粧ボーダーなどの用石が新小松石、白河石（黒目）、芦野石であったこと。
③手加工制作の石工の確保が比較的容易で、思い通りの現場制作が可能であった。
B．1981年以降
①時代が移り、限られた短い工期に合わせ、石工の拘束期間を短縮して、より仕上がり精度の良い制作を遂げる必要に迫られるようになった。従って、素材の石は出来る限り工場加工とし、現場制作では、工場加工された石に最小限の手を加えることによって、完成後はあたかも手づくりで制作されたような雰囲気を生むことが必要となった。
②切石積、化粧ボーダーなどの素材の石が、安山岩系の新小松石、白河石（黒目）、芦野石などの中硬石・軟石から、硬石の花崗岩、ラステンバーグその他に変わった。
　以上、Aは時代と共にBに変わった。
　B①に説明を加えると、素材の石を出来得る限り工場加工とし、現場制作では、工場加工の素材に最小限の手を加えることによって完成後はあたかも手づくりによって制作されたような雰囲気を生ませる為には、制作図を細部に恒って作成する必要に迫られた。
　1991年11月に完成した三番町KSビルの制作図に見られる、克明にして枚数の多さは、それを語っていよう。

制作設計・実施制作図までは、制作者の私自らの手で創られる。出来上がった実施制作図のチェック、素材加工の為の工作図作成と手配、制作員の割り振り、工程計画などは、この面でベテランの高橋久身君が、10年程前から担当している。
　1987年　工学院大学八王子校舎3号館　外空間
　1989年　日野自動車工業本社「日野台の杜」改修
　1991年　スコープ本社 GLASS HOUSE　外空間
　1991年　三番町 KS ビル　外空間
　以上の4つの外空間は彼の担当によったものである。
　各々の外空間は、高橋久身君によって制作員としての石工頭、手元が指定され、材料の現場搬入、下ごしらえ、組立、据付といった順序で進められた。

Prologue

More than two decade, I have been impose discipline on myself to design urban exterior spaces, working within the constraints of the materials and the construction techniques, trying to get close to a certain spatial experience to build fabulous spaces. The spaces thus created were given the name 'Gai-Kukan' by the late Ryuichi Hamaguchi.

In the years since then I have created a number of 'Gai-Kukan' in various cities. The following five spaces are best reflected that my original purpose which creating 'spaces that are unified with a road'. Namely, seemingly the road swells, forming a space that embraces the people there comfortable, using the surrounding building facades as an effective background.

1971 Keio Plaza Hotel, Street No.4, 'Gai-Kukan'
1978 Togin Kurita Bldg. 'Gai-Kukan'
1980 NTT Hiroshima Niho Bldg. 'Gai-Kukan'
1991 SCOPE Main Office GLASS HOUSE 'Gai-Kukan'
1991 Sanban-cho KS Bldg. 'Gai-Kukan'

Each of spaces is a represent of formed shapes and processing materials. By the spirits of traditional Japanese people who loving nature. I am confident that these spaces contain within them the power to break through the cool and impersonal society found in urban spaces and modern man himself.

What I aimed to create was a powerful 'Gai-Kukan'? In order to the question I worked to combine two concepts: 'natural nature' and 'intentionally bold nature'. 'Natural nature' would be something like water flowing quietly and river cobbles, or soft, deciduous trees that swaying for the wind. On the other hand 'intentionally bold nature' might be expressed by pavement or stonework that is geometrically designed or that shapes has a sense of direction, or that aspect of jutting out of rushing water.

In addition to the above five spaces, there are some others I have built within the confines of a building compound. Examples of these are as follows:

'Gai-Kukan'
1971 Keio Plaza 7th Floor Roof Garden
1972 Hino Motors, Main Office, 'Forest of Hinodai'
1972 Hotel Grand Palace
1973 Oji Sports Garden (Razed)
1974 Akabeko Restaurant in Yokohama (Razed)
1974 Konishi Brewing Company Tokyo Branch Office (Razed)
1975 Konishi Brewing Company Main Office VIP Room
1975 Tokyo Bank Aoyama Retreat (Club House)
1976 Wakura Hot Spring Resort Annex "Sun-Kagaya"
1977 Sapporo 'Atsubetsu Commercial Center' (Partly Razed)
1986 Ito-Yokado Group Training Center in Tama
1986 Kogakuin University Hachioji Campus, Bldgs. No.5 to 11
1987 Kogakuin University Hachioji Campus, Bldg. No.3 Residences

1972 Matsukata Garden
1976 Kodera Garden
1987 Mizuno Garden
1977 Guesthouse of Hadazade's (Iranian) Wall Fountain

Each project listed above gave me ideas of spatial design, process and details for my five masterpieces mentioned at the beginning in the design process.

Stone as a material for "the internally bold nature"

In designing and creating details for the above mentioned external spaces many hints are taken from the creation of the mentioned five spaces afore. For instance using stone as a material to create 'intentionally bold nature'. In order to produce a powerful and thought provoking 'Gai-Kukan' that can reform the cold, impersonal space created by the glass, steel and concrete architecture of modern cities, it is necessary that an equally strong material yet one with different characteristics.

In the early stage in 1965-80

(1) Staircase and paved floor

Treads and risers ... granite Inada-ishi (hard stone)
Flag stone...granite
Flag stone...granite Enzan-ishi (hard stone)
Small paving stone...granite Fukushima-ishi (hard stone)
Small paving stone...granite stone produced in China

(2) Stone walls (durable stone walls)

Cut stone...andesite Shinkomatsu-ishi (medium hard stone)
         andesite Shirakawa-ishi Kurome(medium hard stone)
Kenchi-ishi...andesite Shinkomatsu-ishi (medium hard stone)
         andesite Akagima-ishi (medium hard stone)
Stone walls (Ornamental stone walls)

Cut stone...Amphibole, granite, pyroxene andesite,
         Ashino-ishi (soft stone)

(3) Cut stones for retaining wall and ornamental edging.
   Shinkomatsu-ishi
   Ashino-ishi
River cobbles and fragments...Shinkomatsu-ishi
                              Ashino-ishi
Stone work and flagments...Shinkomatsu-ishi
                           Ashino-ishi

(4) Single stone...Shinkomatsu-ishi
(5) Cast concrete or finished with prodding stone to make a

contrast with stone composition.

1981-91

(1) Steps, floor paving...granite Inada-ishi
    Cap stones on the cast concrete wall, or finished with prodding granite 'Inada-ishi'

(2) Stone walls cut stone...Granite, Rustenberg
    Kenchi-ishi...andesite Shinkomatsu-ishi

(3) Ornamental cut edging stone...Granite, Rustenburg
    River cobble and cut stone...Granite, Rustenburg.
    Retaining edging stone...Long cut 3~5m Ashino-ishi.

(4) Single stone...andesite Shinkomatsu-ishi

(5) Others: Cast concrete, finished with prodding, etc.

Change in stone selection after 1981:
Architectural styles changed during that period and a lot more stones was being used to finish the buildings I was working with. I needed materials to bear the same hard situations.

The price of Shinkomatsu-ishi rose steeply and, with regards to concepts of my work, I found a problem of durability with Ashino-ishi.

On design process

If you look carefully at the drawings and photographs you will notice a considerable difference in the atmosphere of my 'Gai-Kukan' as they exist between A. 1971-80 and B. 1981-91. This change is attributable to a number of factors as follows:

A. 1971-80
① It was possible to find stone masons who were capable of doing hand work and it was also possible to set up workshops on site and carry out production in the field, fabricating pieces as needed.
② I used a lot of soft andesite type stones like Shinkomatsu-ishi, Shirakawa-ishi (Kurome) and Ashino-ishi, for cut stonewalls, retaining edging, and ornamental stone edging. These were suited my purpose at that time.
③ Skilled stonemasons were relatively easy to find, and that made it possible to design on site as imagined.

B. Post 1981
① As the time has changed and it has now become necessary to keep the mason's time on as tight a schedule possible and at the same time produce a more highly finished product. This has caused the work to shift from the site to factories with on site finish work keep to a minimum.
② The shift of stone materials such as cut stone walls, ornamental edging and others was made away from soft and medium andesite-type stones like Shinkomatsu-ishi, Shirakawa-ishi (Kurome), Ashino-ishi, andesite to harder stones like granite and Rustenburg.

This is how the era A changed to era B with time.

In addition to B ①, in order to get the workers in the factory to produce work that would look like hand finished stone and yet require little on site finish look, it became necessary to produce highly detailed drawings for production.

The KS Bldg. at Sanban-cho in November 1991 and its enviably great number of detailed drawings tells it all.

The design and production of the drawings has been done by myself. For the last ten years, the checking of these plans, preparation of working drawing, arrangements for construction, allotment of workers and other business details have been handled by Hisami Takahashi, an expert in this field. Mr. Takahashi was responsible for the following 'Gai-Kukan's:

1987 Kogakuin University Hachioji Campus Bldg. No.3 'Gai-Kukan'
1989 Hino Motors, Main Office, 'Forest of Hinodai', Renovation
1991 SCOPE Main Office GLASS HOUSE 'Gai-Kukan'
1991 Sanban-cho KS Bldg. 'Gai-Kukan'

京王プラザホテル［北館］4号街路 外空間 1971
Keio Plaza Hotel (North Wing.), Street No.4 *Gai-Kukan* 1971

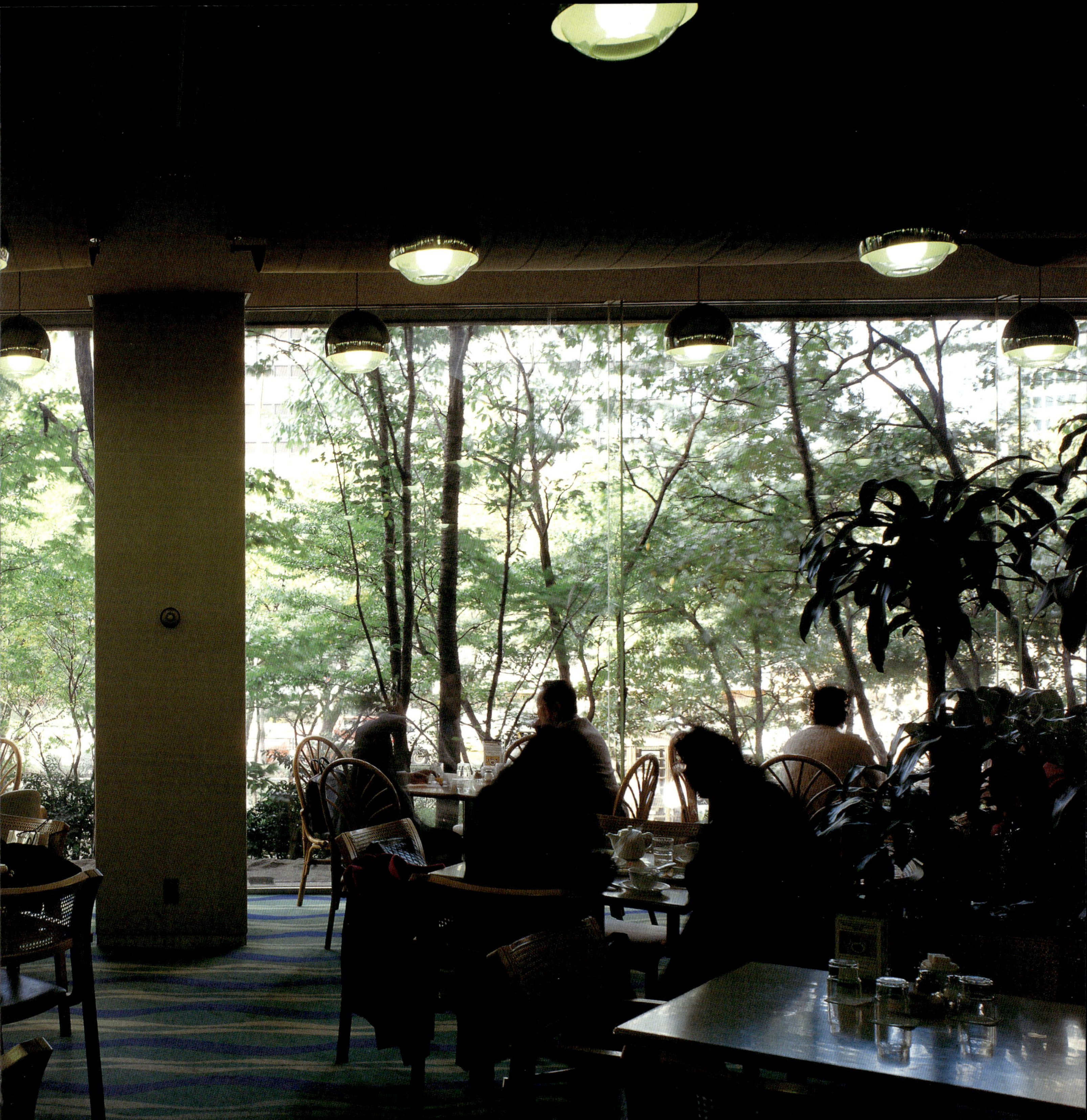

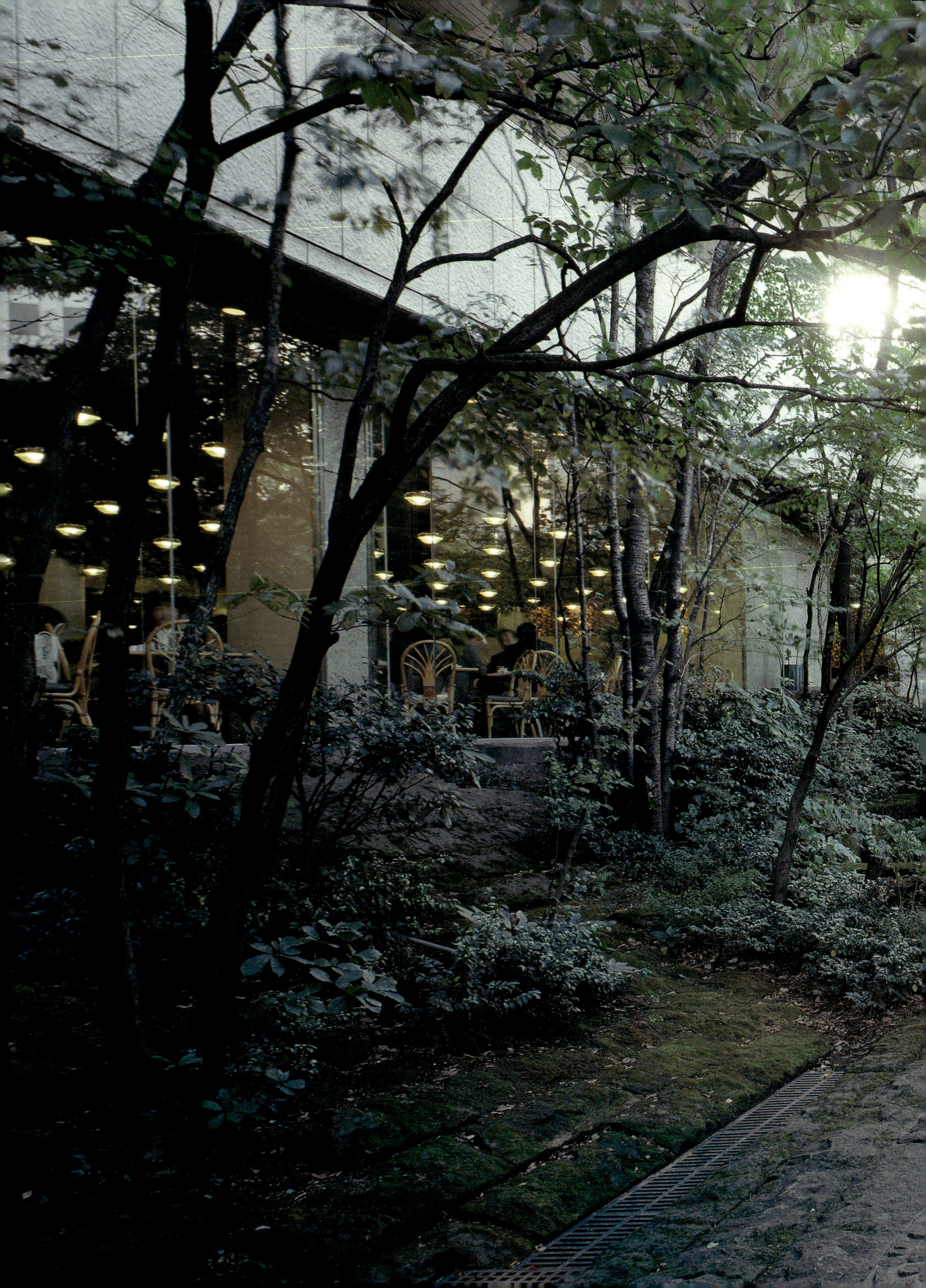

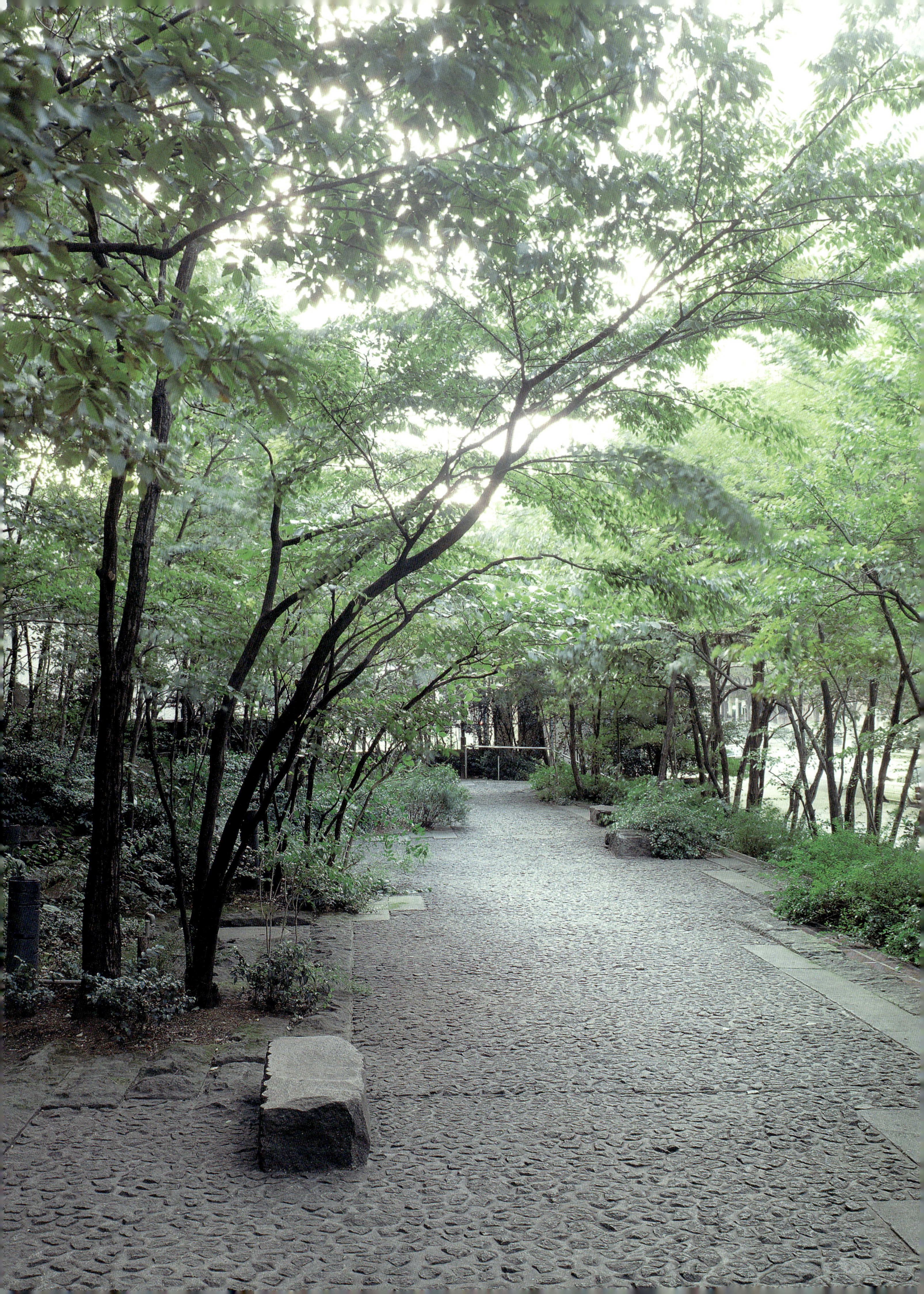

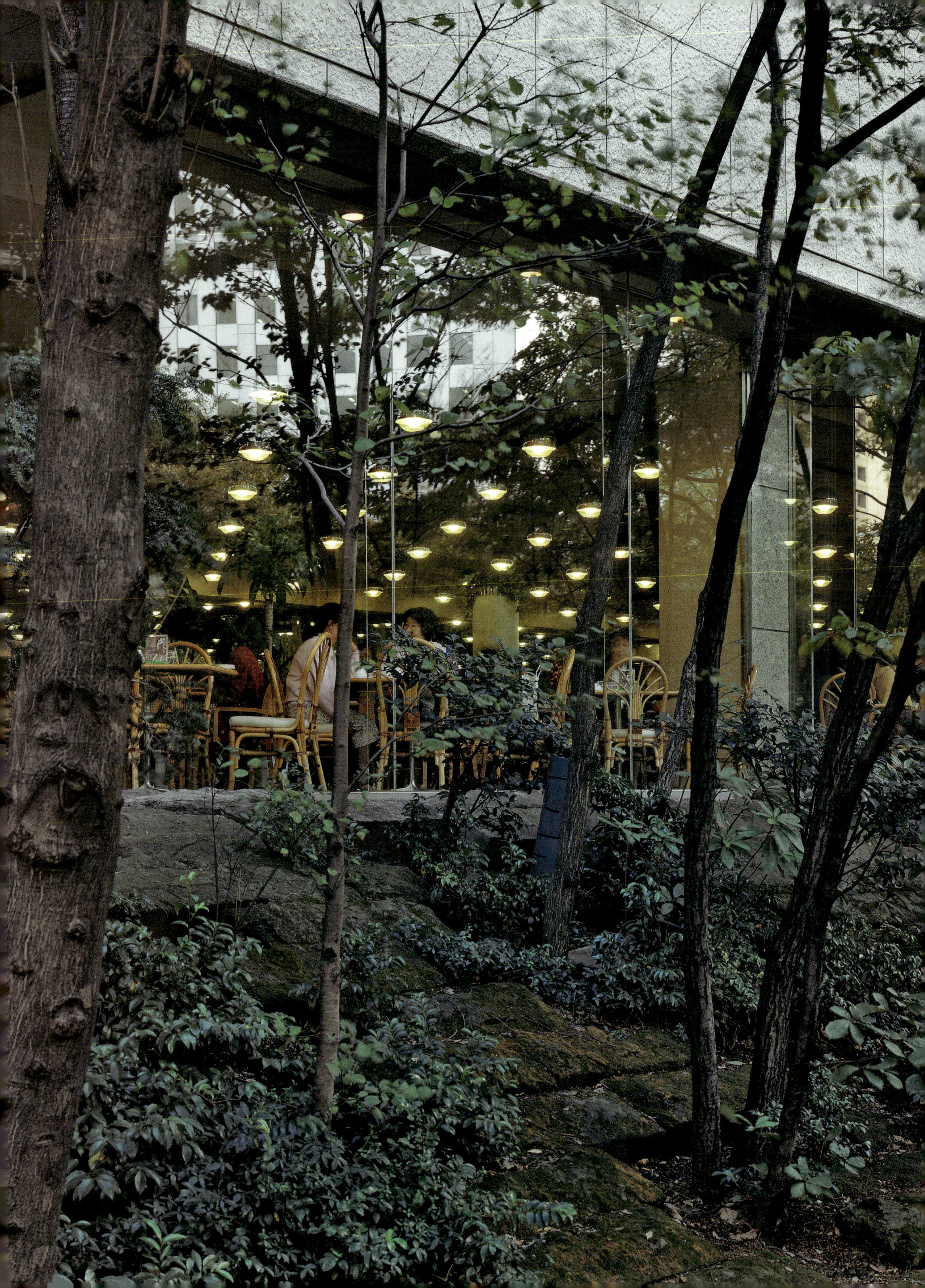

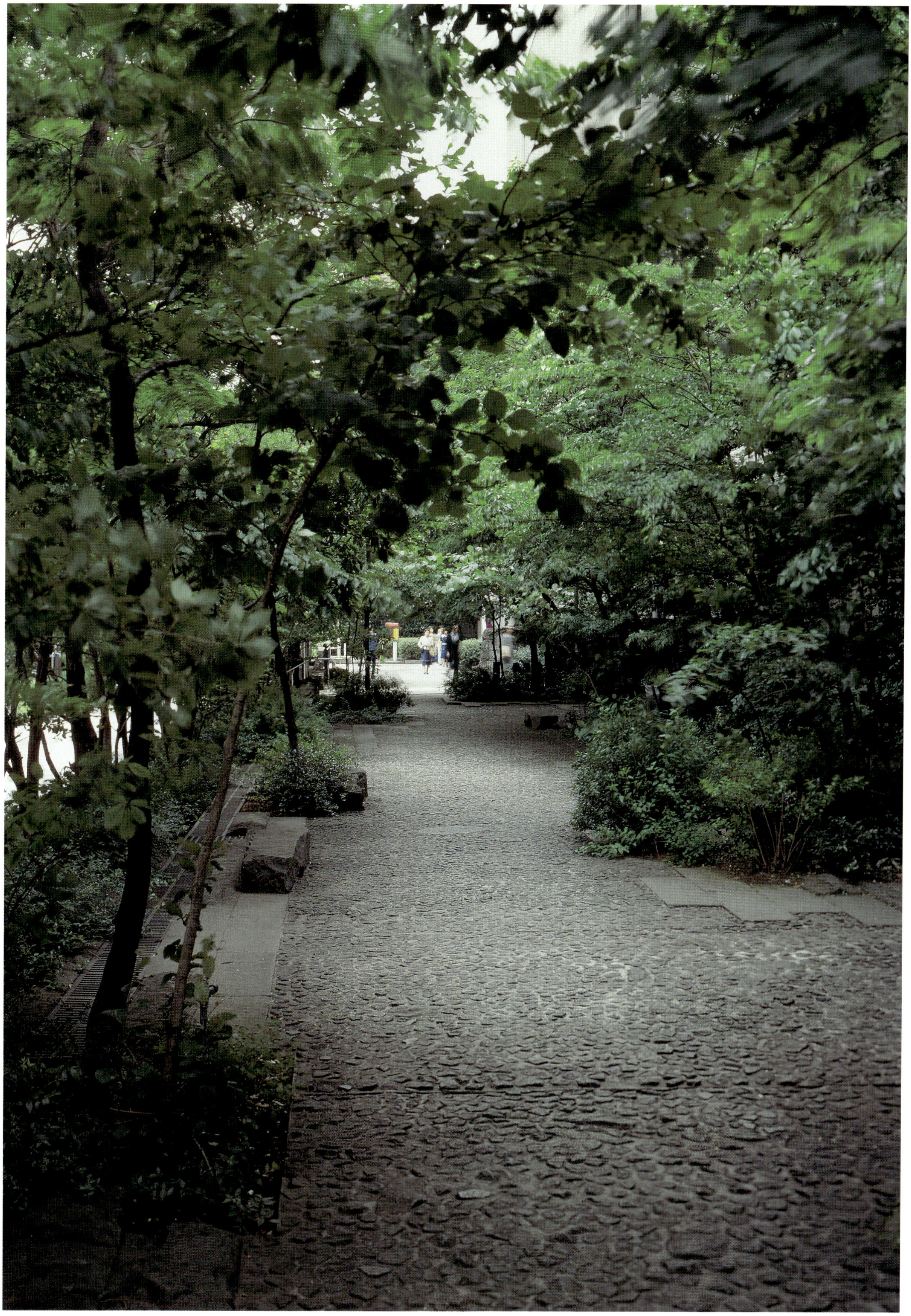

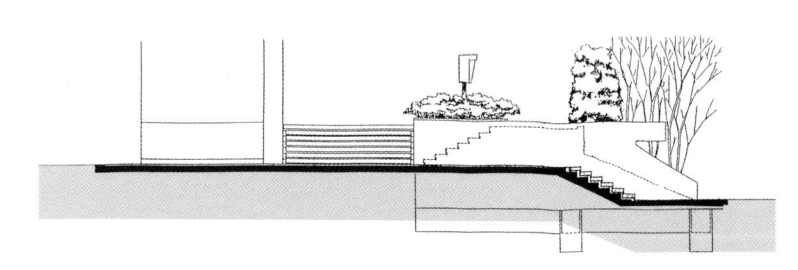

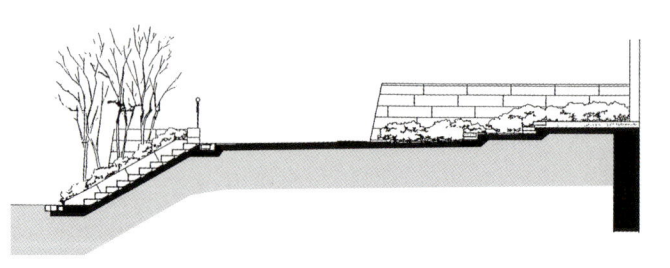

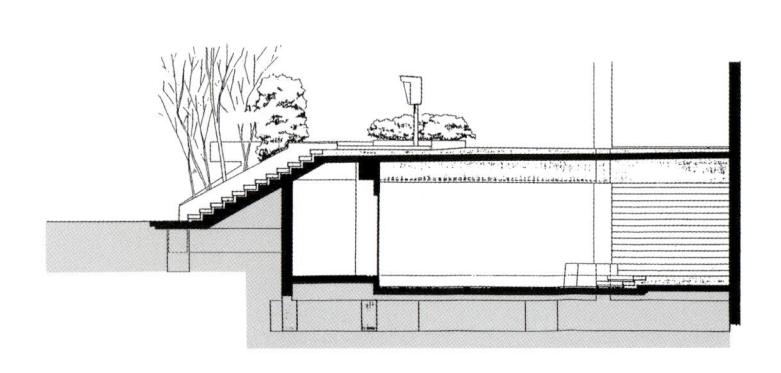

断面図 1/200

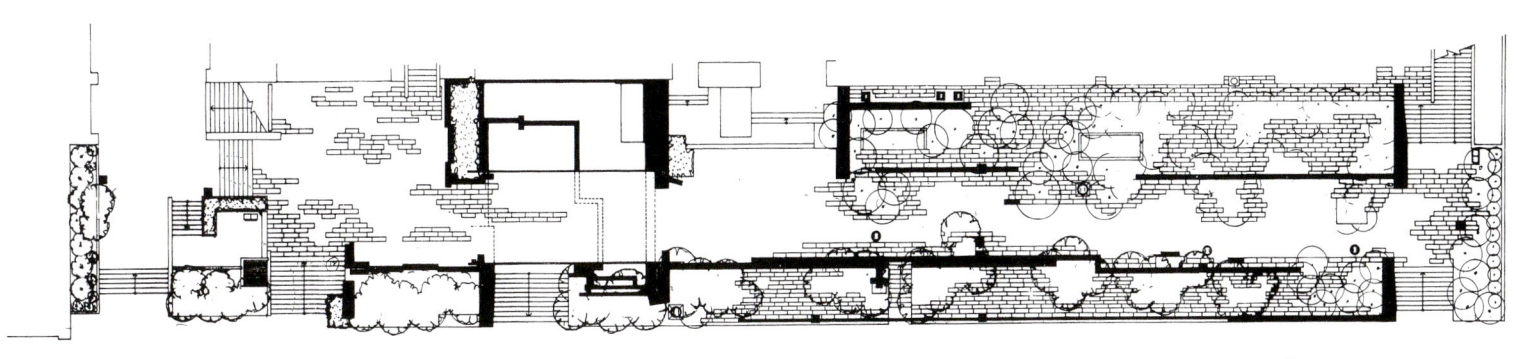

1階平面図 1/500

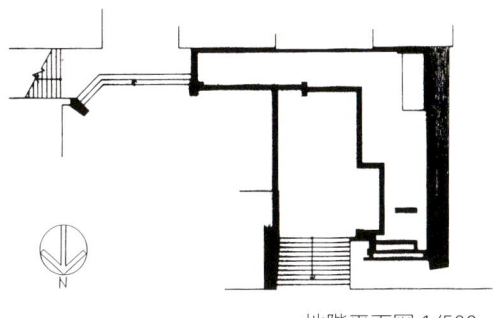

地階平面図 1/500

27

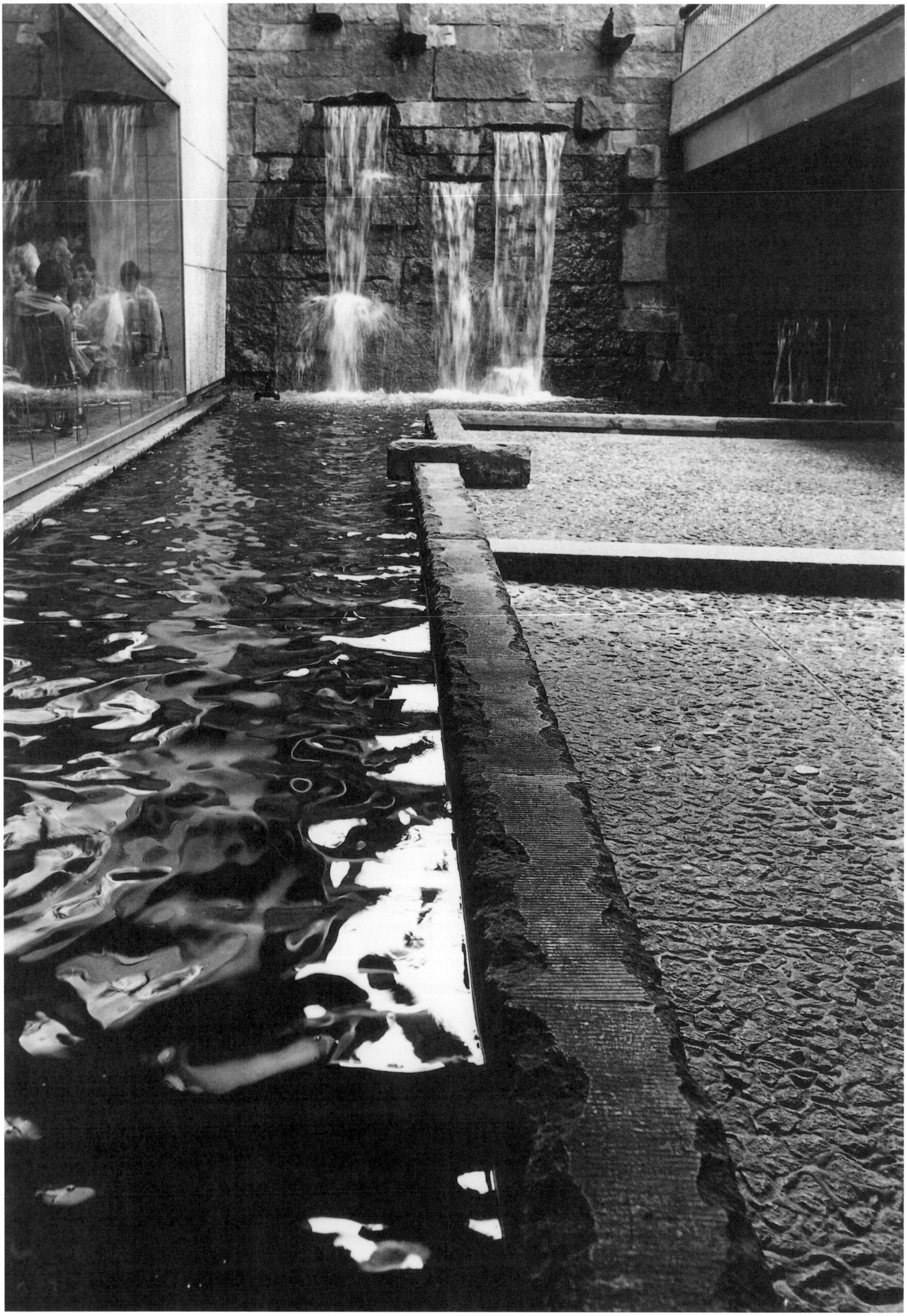

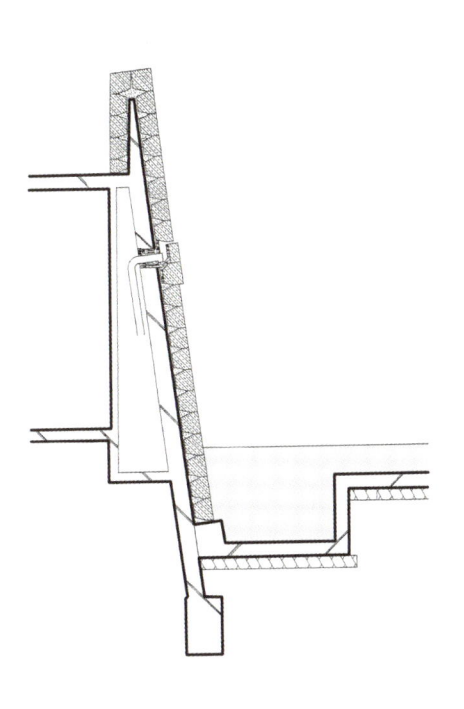

断面詳細図 1/100

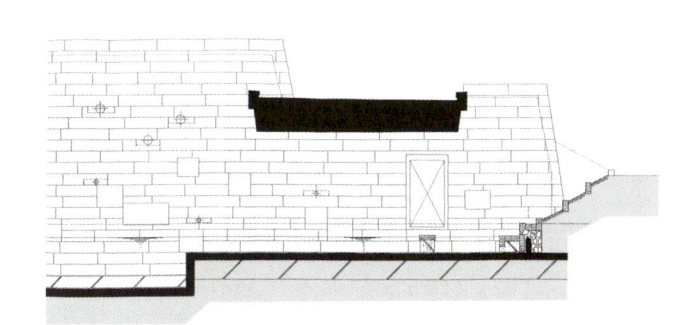

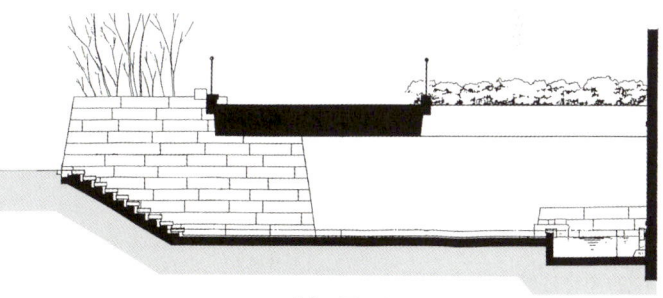

断面図 1/200

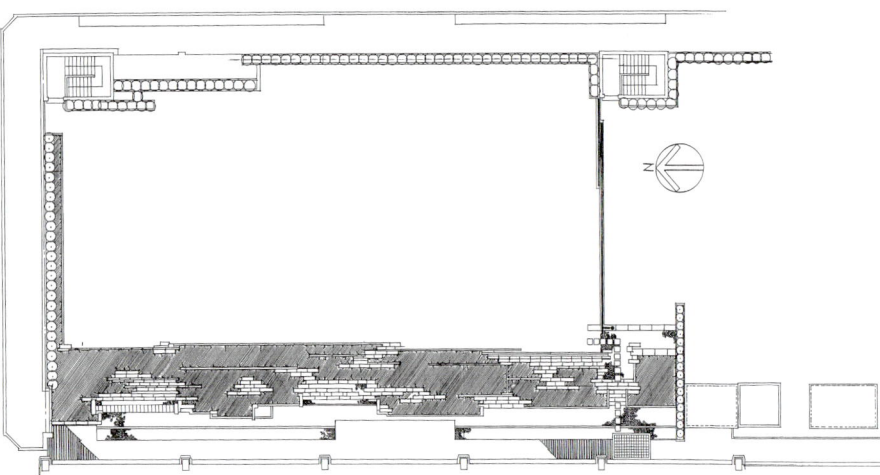

7F屋上庭園平面図 1/600

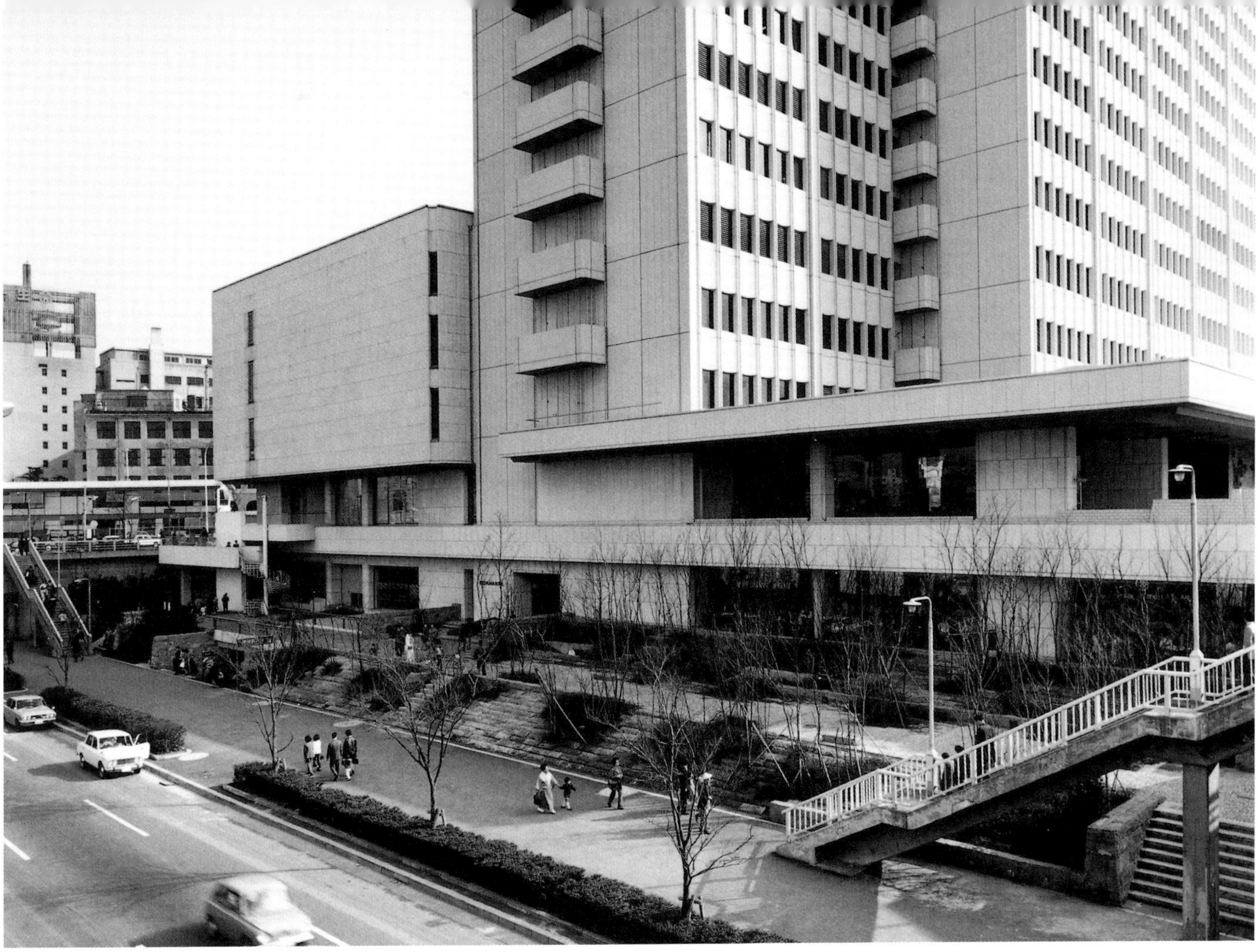

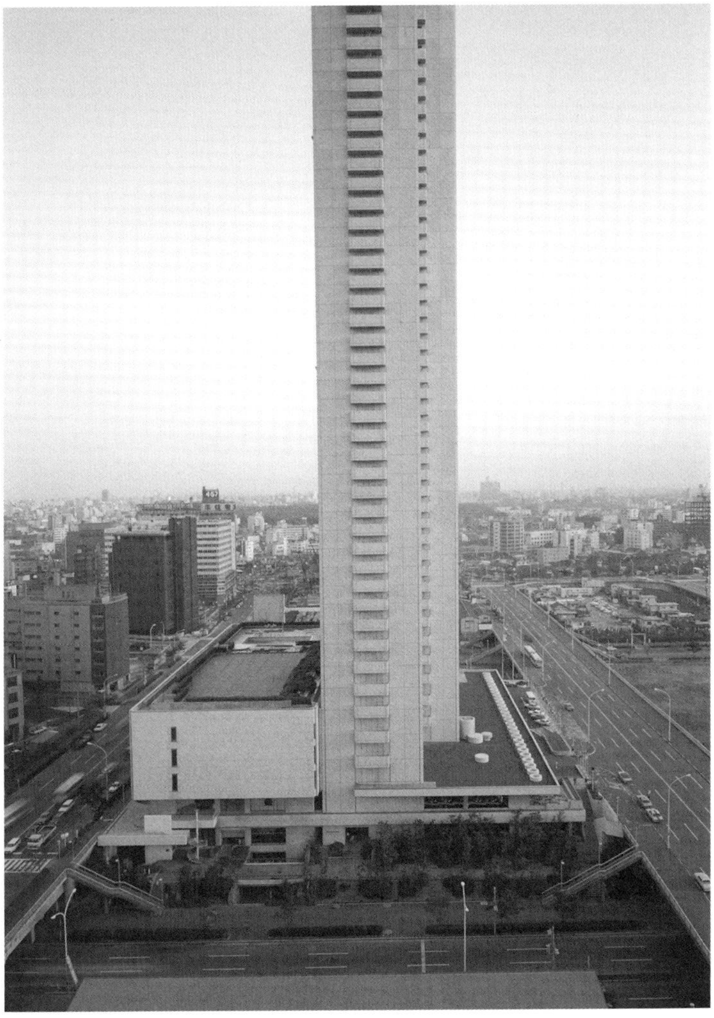

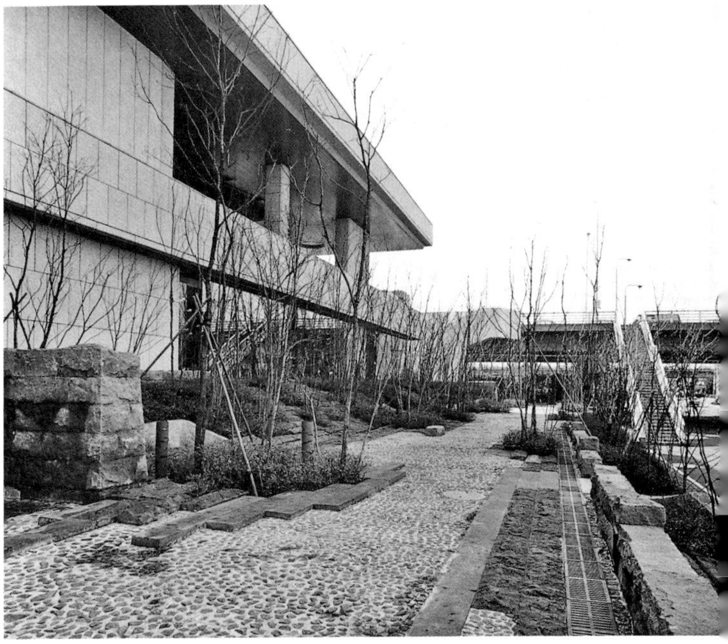

京王プラザホテル［北館］4号街路 外空間 1971

「京王プラザホテル（北館）」は、旧淀橋浄水場跡地に出来た副都心地区のビル建設第1号として、1971年3月に完成した。

北館唯一の緑のあるこの外空間は、北館の北側に並行して走る4号街路、4号街路の東側に層を変えて交差する9号街路、4号街路の西側に層を変えて交差する10号街路との間の100m。北側外壁から4号街路境界まで15mの長さの内部に、街区に規制された公共提供の幅6mの歩行者通路としての平坦地を含む面積1,500㎡の細長い空間で、20数年前に目指した、道がふくれ、まわりの建物のファサードを効果的な背景として人びとを快適に抱き込むような空間に近付き得たと思われる空間第1号で、1971年6月完成した。

このような人間集団・大衆を相手に展開されるべき空間では、それぞれに対して"いたわり"という後始末を大きな視点に展開すべきであろう。言葉を換えれば、人間のアイレベルの範囲での空間展開が望ましいということであろうか。

人間のアイレベルで建物と外部空間との間に安定感を保たせようとするためには、建物の外壁と外部空間、部分及び類似部と外部空間との接点、可視障壁部と外部空間との空間接続と展開が要素として取り上げられよう。

しかし、それはそれとして、建物の外部空間に介在する人びとが、超高層ビル故に受ける威圧感、微気象故に感じる違和感からの解放が空間展開の大きなテーマであった。

このような細長い空間は、空間をいくつかに分けて展開し、互いに演出し合わせることによって、より効果あらしめることが定石であろう。

モチーフは、この地が旧淀橋浄水場であったことを含んで「この土地に眠る歴史」と「かつての自然の風物」などとした。

以上をモチーフに抽象表現することとし「自然な自然」と「意図された逞しい自然」とが組み合った空間構成と展開を行うことに決めた。

オープンカット工法によるビル建設跡地の埋戻しを粘土混じりの山砂とし、街路境界から北側外壁まで15mの狭い敷地幅に遠近法によって拡がりを持たせる為と、埋戻しによって起こるであろう不等沈下を避ける為に、街区に規制された幅6mの、公共に提供の歩行者通路としての平坦地を残し、1層あるいは2層の斜面とした。

その上に、土砂を押さえるスラブの上に吸水性の大きい芦野石（角閃石・石英・両輝安山岩）を斜めに岩盤状に貼り、岩盤の上に建つ超高層ビルとして安定感を持たせ、将来環境に応じて発生する植物の住みつきに可能性を持たせ、岩盤の処々に雑木林と風のそよぎをよく感じる事ができる潅木を配植し、超高層ビルとの視覚的にナイーブなアイレベルでのつながりを計った。

他方、Y20-Y24（166P参照）に、建築計画の上で生まれた5～6mの落差を利用して、あたかも地下水の噴出によって生じたかの様態の滝を新小松石の石積を利用して落とし、住みつくべくして住みついた自然の様態を展開した。

素材と造形手法を厳しく限定、自らが自らに課した規律の中で、一定の空間体験に近づこうとしたそれらは、かつて私の体験した自然で、武蔵野でもある。

この空間に介在する人びとは、この新しい自然に果たして慰謝されているだろうか。

風土が日本文化の決定的基盤であるように、日本の都市の外部空間も又、風土が決定的基盤的基盤であろうとした若き日の空間展開の一つである。

```
1971年                          1983年
用石の種類                       石 積    新小松石間知石
石 積       新小松石切石          階段袖石  新小松石
化粧ボーダー    〃              階段石    既存転用稲田石
斜面敷石     芦野石切石          一部縁石   〃
縁石支縁石笠石   〃              土 留    鉄筋コンクリート打放
一部石積      〃                水の様態   湧水、流れ、滝
階段石      稲田石切石          自動灌水   スプレーノズル式
床 石        〃

1982年
床敷石      中国産みかげ石
床飾石      印度産みかげ石ニューインペ
```

※詳細図はディテール図面編の166P参照

京王プラザホテル7F屋上庭園

　この下の階、6階は大宴会場である。従って、この屋上庭園の床はロングスパン構造のスラブであるから、荷重に制限があった。

　現在、石積、それに堀、潅木が植えつぶされている部分と屋上の四囲を取り囲む生垣の部分が荷重1t、他は、クリンカータイル貼りの部分が荷重100kgである。

　このような空間の処理は、庭園として屋上に在ることを感じさせないといった定石と、どの方向からの眺めも奥行きの狭さを感じさせない技術が必要であろう。

　実施の手段として、空間に絶えず視的緊張感を誘う対象としての水の動きを中心に構成展開を行った。

　建物側からの眺めの石積の平面及び立面構成は、日野自動車工業「日野台の杜」の石積に近い構成と同じ手法を用いている。

1971年
用石の種類
石　積　　　　　　芦野石
化粧ボーダー石　　　〃
床石、敷石　　　　　〃
飛　石　　　　　　　〃
階段石　　　　　　　〃
水の表態　　　　　流れ、滝、堀
自動潅水設備　　　スプレーノズル式

※詳細図はディテール図面編の167P参照

Keio Plaza Hotel (North Wing.), Street No.4 '*Gai-Kukan*' 1971

The North Wing of the 'Keio Plaza Hotel' was completed in March 1971 as the first building in the newly developed city center, on the lot where the Yodobashi water purification plant used to be.

This '*Gai-Kukan*', garnished by greenery and unique to the North Wing is 100m in length and surrounded by three streets; Route 4 runs parallel to the north side of the North Building, Route 9, which intersects with Route 4 on a different level on the east side, and Route 10, which also intersects with Route 4 on a different level on the west side. Inside the 15m wide, 1,500㎡ narrow strip of space between the north side of the North Wing and Route 4, a 6m wide flat terrain that is also a public footpath is included. This path complies with city block regulations. In June 1971, this became the first spatial project to achieve my long-held aim of over 20 years to design a space that embraces people comfortably in a road swells, using the façade of the surrounding buildings as effective background.

A space like this that is intended for a wide range of user groups should be planned with a broad view of aftercare and consideration. In other words, the views from eye-level should be a major focus of the design.

In order to keep a sense of stability at eye-level between the exterior space and buildings, space connection and design would be important elements; between the exterior space and the external walls of the building, the point of contact with parts or similar parts and the exterior space, visible barriers.

However, having said that, the primary theme for the spatial design is to liberate the people who interact with the exterior space of the building from the overwhelming pressure of the high-rise buildings and a sense of incongruity due to the microclimate. A standard move for such a narrow space is to divide it into sections and design each to enhance the other thus increasing the effect.

Inspired by the site that used to be a water purification plant, I chose 'Sleeping history of the land' and 'Past natural scenery' as my motifs and expressed them in the combined space configuration as 'natural nature' and 'intentional, bold nature.'

A mixture of sand and clay was used for the backfill on the open-cut construction site. Leaving flat land for the 6m wide public footpath required by local regulations, I created one or two terraced slopes to give a false perspective to broaden the view of the narrow 15m wide site between the north side of the North Wing and the street, and also to avoid the risk of uneven sinking of the filled site.
Compressing slabs are laid on the sand layer, and Ashino-ishi, a stone with high water absorbing properties, consisting of amphibole, quartz, and two-pyroxene andesite, on top of that, slanted to look like bedrock and thus giving a look of stability to the high rise buildings built on it. In order to give plants the chance to grow in the future in accordance with the new environment some deciduous trees and shrubs that would move nicely with the breeze were planted here and there on the bedrock, and designed to give a visually gentle connection to the high-rising buildings at eye-level.

In another place, labeled Y20-Y24 (refer p166), a waterfall was created using Shinkomatsu-ishi stone wall to make the most of the 5-6m level difference that occurred because of the architectural design, and made to look like a natural spring, leaving plants to grow naturally.

Within my own strict rules and use of carefully chosen methods and materials, the attempt to get close to a certain spatial experience was actually re-experiencing a scene in nature that I had previously seen in Musashino district.

I wonder whether the people who interact with this space are comforted by this newly created nature.

As the environment is a decisive base for Japanese culture, so might the Japanese urban exterior site be; a spatial design thought from my youth.

1971
List of stones
Stone walls: Shinkomatsu-ishi, cut.
Ornamental edging stone: same as above.
Paving on the slope: Ashino-ishi, cut.
Stone curb and coping: same as above.
Part of stone walls: same as above.
Stone treads: Inada-ishi, cut.
Paving: same as above.

1982
Stone paving: Chinese granite.
Decorative stone paving: Indian granite 'New Imperial'.

1983
Stone walls: Shinkomatsu-ishi, Kenchi-rocking-wall.
Stone piers for steps: Shinkomatsu-ishi.
Stone treads: existing Inada-ishi.
Part of stone curbs: same as above.
Retaining wall: Cast concrete with steel bars.
Water features: Spring water, stream, and waterfall.
Automatic irrigation system: Spray nozzle.

※ details 166p

Keio Plaza Hotel. 7th floor. Roof-top garden

The 6th floor, one floor is below this. The area is a large banquet hall.
Therefore the floor of the roof-top garden is made of long span slabs and has limited capacity on carrying weight.

The areas of stone wall, The pond, the shrubs, and the fence surrounding the roof top weigh one ton altogether. The rest. the clinker-tiled area. weight 100kg.

For the processing of this kind of space, one needs the usual technique of not letting people feel that who are living on top of a building, or that the garden has rather limited depth in all directions.

In the actual composition, arrangement was made to place a movement of water in the center, which always visually stimulates people.

In composing the horizontal and vertical planes of the masonry as seen from the building side, a similar technique was used to the case of stone wall in Hino Motors, Main Office, 'Forest of Hinodai'.

1971
List of Stones
Stone walls: Ashino-shi.
Ornamental edging stone: same as above.
Stone paving: same as above.
Tobi-ishi: same as above.
Stone treads: same as above.
Water features: Stream, waterfall and moat.
Automatic irrigation system: Spray nozzle.

※ details 167p

日野自動車工業本社「日野台の杜」外空間 1972
Hino Motors, Main Office, 'Forest of Hinodai' *'Gai-Kukan'* 1972

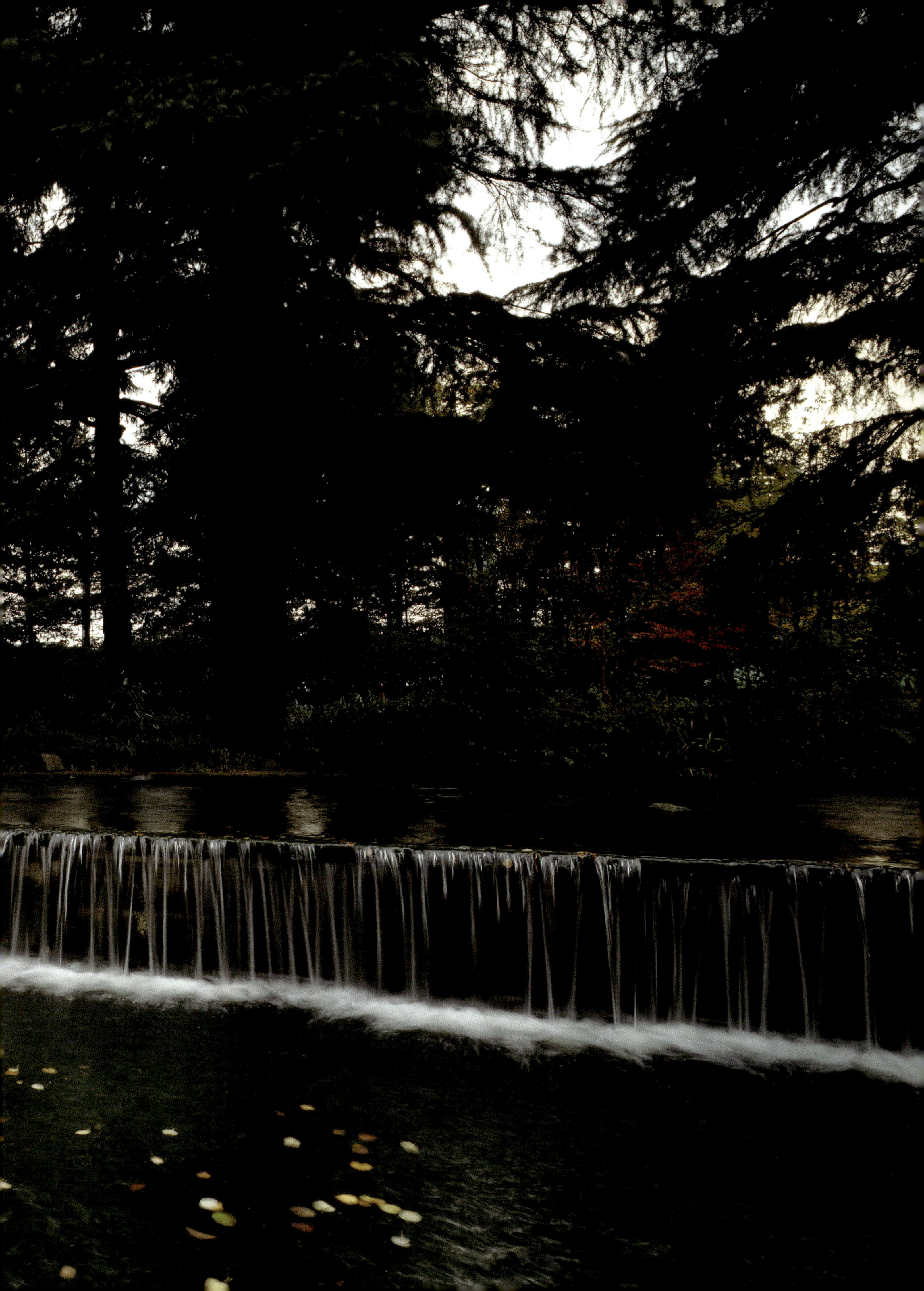

38

日野自動車工業本社「日野台の杜」外空間 1972

　我が国には珍しい鉄骨現し型6階建ての日野自動車工業本社屋ファサード（南向き）から国道20号線境界まで40～50m、面積約6,500㎡（社屋前車路含む）が「日野台の杜」（外空間）で、1972年完成した図のような形の空間である。

　本社屋ファサードのエントランス前にひときわ高く聳えるヒマラヤスギの林と、この林を中に、サワラを混じえた落葉樹林（雑木林）が左右に展開している空間がある。

　これは1970年、新社屋完成を機に、三代社長・松方正信氏の英知によって生まれた空間で、空間の中心となっているひときわ高く聳え立つヒマラヤスギの林（群植）は、創始者の松方五郎氏が創始時に植栽された記念樹林である。今回、このヒマラヤスギの林を会社の表徴として遺し、精神空間を創り出そうという松方社長の発意によっている。

　日野自動車工業の立地する日野市は、20数年程前まで湧水が豊富で、他方、古代人の住居跡がしばしば発見・発掘され、日野自工の近くにも七つ塚と名付けられた古墳があった。（現在は定かではない）

　こうした事実から、水を求めこの地に定住した先人、厳しい気象の中で自然に抗し生き、そして死んで往った長い列を永遠に忘れないようにとの配慮から、象徴して遺されたヒマラヤスギの林を縫ってカスケードを流し、幅広い瀑布を堀に落とした。そして武蔵野の表徴であった雑木林が、これらを中心に左と右に展開する。

　雑木林の一隅に建てられた碑に、「武蔵野の面影を後に遺そうと、本社屋完成を機に計画された庭園がここに完成した。この庭園には、永遠に変わることのない武蔵野の面影があり、日本の心、日野自動車の心がある。嘗て先人の遺してくれたヒマラヤスギは、今、力強い枝振りを誇り乍ら、我々にその心を語りかけてくれる。我々も又……」とある。ともすれば忘れがちな心構えを、この碑が教え、林が語る。

　松方社長と共にこの庭園を創り出した私は、訪れる度に手のとどかない処へ離れていく木々に、流れ去った歳月を想い、朝露、葉ずれと共に生誕する神々を思う。

　この空間は1982年、創立40周年記念碑設置、これに伴う林の一部改修。庭園は「日野台の杜」と改名。

　1989～1990年、「日野台の杜」全域保存の為の改修と保修。

　1971年から72年の間に、3つの空間が相前後して完成した。

① 1971年　京王プラザホテル4号街路　外空間
② 1972年　日野自動車工業本社「日野台の杜」（外空間）
③ 1972年　ホテルグランドパレス　外空間

　これらの3つの空間創りから、

①から、1978年東銀栗田ビル外空間、1982年NTT広島仁保ビル外空間、1991年三番町KSビルの構成・展開の一部

②から、1982年NTT広島仁保ビル外空間の雑木林の構想、1975年東京銀行青山寮の樹林の構想

③から、1991年スコープ本社GLASS HOUSEの外空間の構想の参考になった。

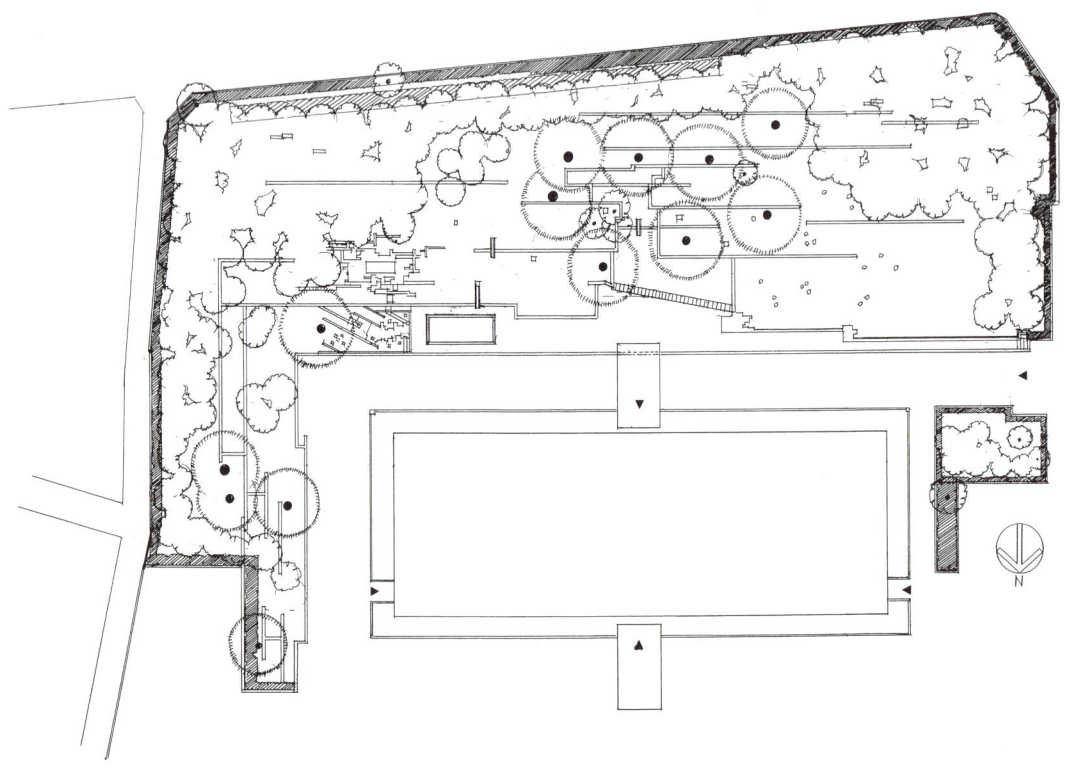

平面図 1/1000

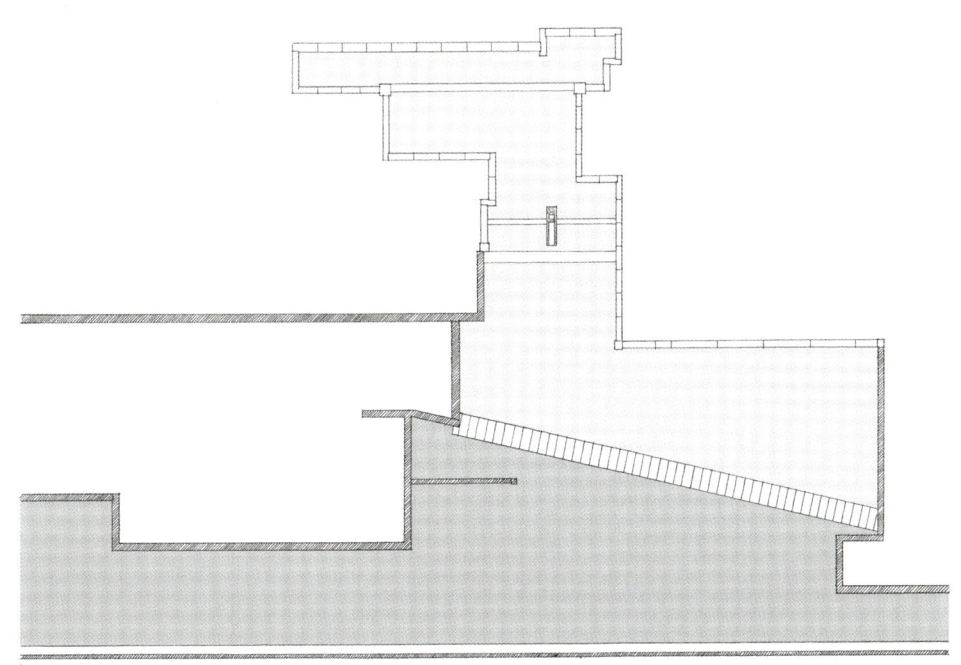

カスケード、滝部 平面図 1/300

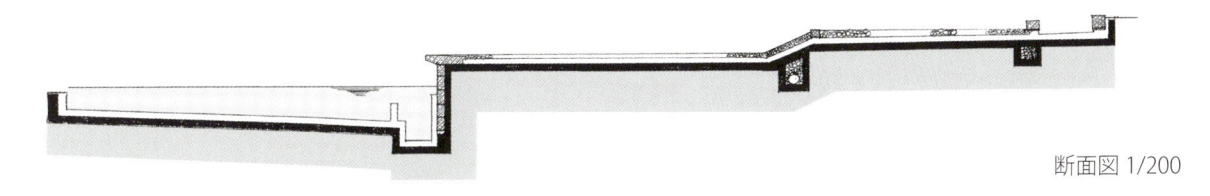

断面図 1/200

Hino Motors, Main Office, 'Forest of Hinodai' *'Gai-Kukan'* 1972

The *'Gai-Kukan'* of 'The Forest of Hinodai' was completed in 1972 in the space between the south facing façade of Hino Motors, Main Office, an exposed steel-frame 6-story building, which is uncommon in this country. And 40~50m to the edge of National Route 20, 6,500 m² in area including the driveway to the office building, and its shape is shown in the drawing.

Towering woods of Himalayan Cedars and their surrounding mixed woods of deciduous trees and Japanese False-Cypress grow on either side of the main office building, in front of the entrance and façade.

This is a space created from the wisdom of Masanobu Matsukata in 1970, the third president of the company, and the remarkably tall woods of Himalayan cedars were planted for commemorate the establishment of the business by Goro Matsukata, the founder. President Matsukata suggested retaining the Himalayan Cedar woods as the symbol of the company, and creating a spiritual space.

Hino City, where Hino Motors, Main Office is based, used to have abundant spring water until about 20 years ago. Moreover, some ancient ruins were discovered and excavated, and there were even some ancient tombs called Nanatsuduka near the Hino factory site. (It's uncertain at present).

In response to the site history, a cascade was built weaving among the symbolic Himalayan Cedar woods and a wide waterfall flows into a moat. This is to permanently preserve the long chain of ancestors' lives, the ancestors that had, in search of water, settled here for generations battling nature in the severe climatic conditions. The characteristic deciduous woods of Musashino plateau spread either side of the water features.

A stone monument placed in one corner of the deciduous woodland reads; 'A garden was built here on the occasion of the completion of the main office building to pass on the image of the Musashino plateau to future generations. This garden possesses the everlasting image of the Musashino plateau, as well as the Japanese spirit and Hino Motor's spirit. The Himalayan Cedars that our predecessors at one time left us are now branching out vigorously, telling us about that spirit. Also we....' It continues. The stone memorial reminds us of easily forgotten attitudes and the woodland narrates.

Each time I visit the garden that I created with President Matsukata, my thoughts go to the passing of time by seeing the trees that have grown out of reach, and I get a feeling of the birth of gods among the rustle of leaves and the morning dew.

In 1982, a commemorative stone monument was erected for the 40th anniversary of the establishment of business, followed by partial improvement of the woodlands on this site. The garden was re-named 'Forest of Hinodai'.

1989-1990 'Forest of Hinodai' were improved and maintenance carried out as part of the preservation of the entire site.

The following three spaces were completed in tandem between 1971 and 1972:
① 1971 Keio Plaza Hotel (North Wing.), Street No.4, *'Gai-Kukan'*.
② 1972 Hino Motors, Main Office 'Forest of Hinodai', *'Gai-Kukan'*.
③ 1972 Hotel Grand Palace, *'Gai-Kukan'*.

The creation of these three spaces led to other creative ideas for the following projects;
From ① *'Gai-Kukan'* for the Togin Kurita Bldg in 1978, *'Gai-Kukan'* for the NTT Hiroshima Niho Bldg, in 1982, Partial composition and development of the Sanban-cho KS Bldg in 1991.
From ② the concept of deciduous tree woods in the *'Gai-Kukan'* for the NTT Hiroshima Niho Bldg, in 1982, the concept of woodland in the Bank of Tokyo Aoyama Retreat, in 1975.
From ③ the *'Gai-Kukan'* plan for the Hotel Grand Palace, a *'Gai-Kukan'* plan for the GLASS HOUSE, the Scope Main Office in 1991.

「日野台の杜」

　ヒマラヤスギの群植を中心に武蔵野の自然に帰って往くことを願って構成・展開された雑木林は、10年後（1982年）の今、見事な武蔵野の雑木林の自然を生成し、林を右と左に分けて流れるカスケードや堀の石の構築群、点在する石との間に調和のとれた美を生み始めている。

　30周年から40周年への会社の歩みと共に、この空間も年というものの協力によって、予測以上の空間生成を遂げたのだ。この成果は、創作し制作した私にとって、大きな喜びである。40周年の今、これまで無名であったこの空間を「日野台の杜」と名付けた。

　この度創業40周年を記念して、役員・従業員一同寄贈の「日野の湧水」（石彫）が「日野台の杜」の部分に加えられた。これを機に「日野の湧水」の空間存在の充足理由を成立させ、これまでの空間との調和を保ち、他方、空間に漂う"詩"に加えて、更に深い風景と人とを結ぶ意味、物と人とを結ぶ意味を生ませることを構想して、ヒマラヤスギの林から「日野の湧水」までの空間に、幾つかの石の造形物を配して一つの風景としてつなぎ、将来この空間に介在する人びととの間に、より深い感動の美を生む空間生成を願った。

　これは、以前から石の構築と点石、新たに加えられた石の造形群が、時というものの協力によって風化し、反自然的（人工）という対比的なものを前提として、辺りのヒマラヤスギの林に代表される自然と相俟って、一層鮮やかな自然を生むと考えるからである。

　この空間「日野台の杜」は、今後会社の歩みと共に、空間存在の意味を人びとに伝え続けながら、名実共に「日野台の杜」を生成することであろう。時の流れと共に"杜"と名付けた創作者の意図が理解されると確信している。これが私の外空間の芸術である。

　これからの「日野台の杜」の生成の道程は、厳しく遠い。この自然の生成を静かに温かく見守ってやってほしい。それだけ多くの深い意味を持つ空間だから。

1982年9月

'Forest of Hinodai'

The woods of deciduous trees around the group of Himalayan Cedars were originally designed to blend in to the nature of Musashino in time. Now, in 1982, after 10 years, it has grown into a great natural deciduous woodland of Musashino, and is creating harmonious beauty among the cascades dividing the woodland in the center, a group of stones in the moat, and stones dotted around.

Together with the company's progress, between the 30th and 40th anniversaries, the landscape has, with the help of time, achieved a space much greater than expected. It is a great pleasure for me, the landscape designer. On this occasion of the 40th anniversary, this landscape that used to be anonymous was named 'Forest of Hinodai'.

In commemoration of the 40th anniversary, the 'Spring Water of Hino', a stone sculpture was presented by all the executives and employees and placed in 'Forest of Hinodai'. Taking this opportunity, I would like to desire that be satisfactorily rationalized what the reason for the 'Spring Water of Hino' s' existence, and retaining harmony with the surrounding landscapes. Furthermore, adding to the 'poetry' in the air, and I expected to give birth to the meaning of connect with people and the landscape, people and materials, deeply, it was designed by placing some scattered stone sculptures in the space between the 'Spring Water of Hino' and the Himalayan cedars , for connecting them as a landscape, I wish for the space to grow to bring even more deeply touching beauty to the people to whom interacting with the site in future.

I think that a group composition of the existing stones and focal point stones, and newly added stones that are weathered over time, on the premise of being in contrast with nature, or 'artificial', coupled with nature such as the Himalayan Cedars, would create an even more vibrant natural landscape.

I believe this landscape, 'Forest of Hinodai,' will progress along with the company while continuing to convey its raison d'être to the people, and will grow as 'Forest of Hino' in name and in reality. I believe that with time, the intentions of the creator named 'Mori' will be understood.

This is my art of 'Gai-Kukan'.

'Forest of Hinodai' will travel a long and difficult road. May this generation of nature be watched over warmly and gently because it is a space that has so many deep meanings.
September 1982

「日野の湧水」

　造型物を既成の空間に持ち込もうとする行為は、その造型物が、空間にそれ自体独立して存在することはあり得ず、必ず他のものとの間に関わり合いを持ち条件付けられるから、造型物の大きさ・形・配置について慎重に検討の要があった。
　その上、40周年という10年間の節目を記念しての役員・従業員一同の寄贈になるものであるから、造型物の空間存在の充足理由を充分に成り立たせることが大きな課題となった。
　この2つについて検討の末、造型物を1970年の役員会の発意によって生まれた構想の空間「日野台の杜」の構成の部分に加え、風景の中に一体化し、その心を後世へ伝え遺すことがよかろうと結論し、現在のような大きさ・形・配置とした。「日野の湧水」は、かつて日野市が東京都下有数の湧水の地であったことから、湧水を形象し、その痕跡を再生・保存し、豊かな湧水を表現することによって、会社の安泰と繁栄を表した。他方、水盤に刻まれた次第に径の拡がる4つの輪は創始以来10年ずつの節目を、最後の輪は50周年に向けての未踏の道程を表し、この部分は磨かれている。さて、水盤を支えているかの様態の頭石に刻み込まれたVの部分は、人びととの間にどんな対話を生むであろう。
　湧水から堀りまで浅く流れる水。石の表を清らかな水が、どこが水際ともわからぬ体で流れる。短い流れだが、何時の日か、人の心をゆするであろう。これも「日野の湧水」である。「日野の湧水」は、時の流れと共にその空間存在の意味を深く人びとに伝え続けて往くであろう。
　価値あるものとは、そのようなものではなかろうか。
"心"は私達の宝である。
1982年9月

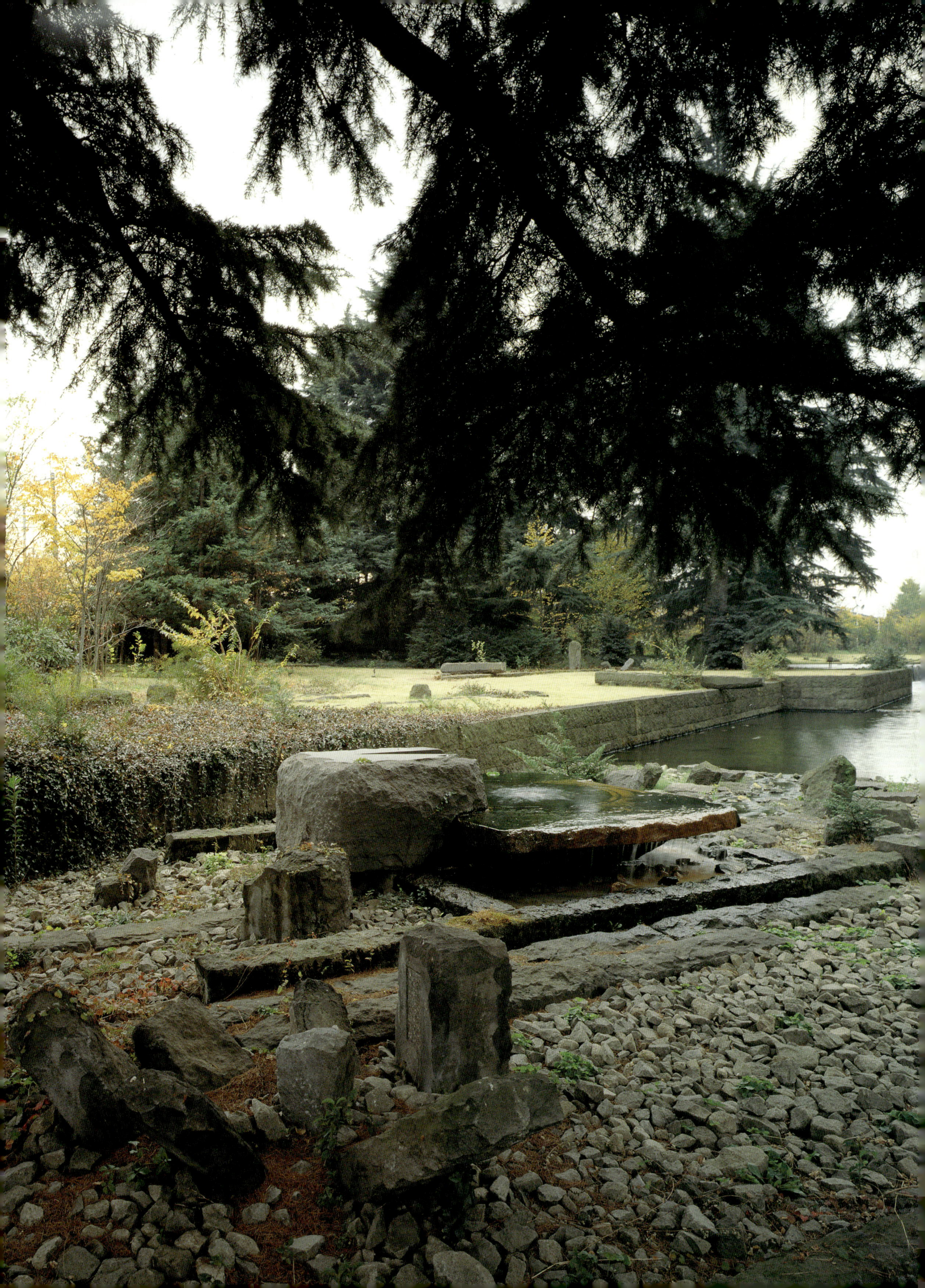

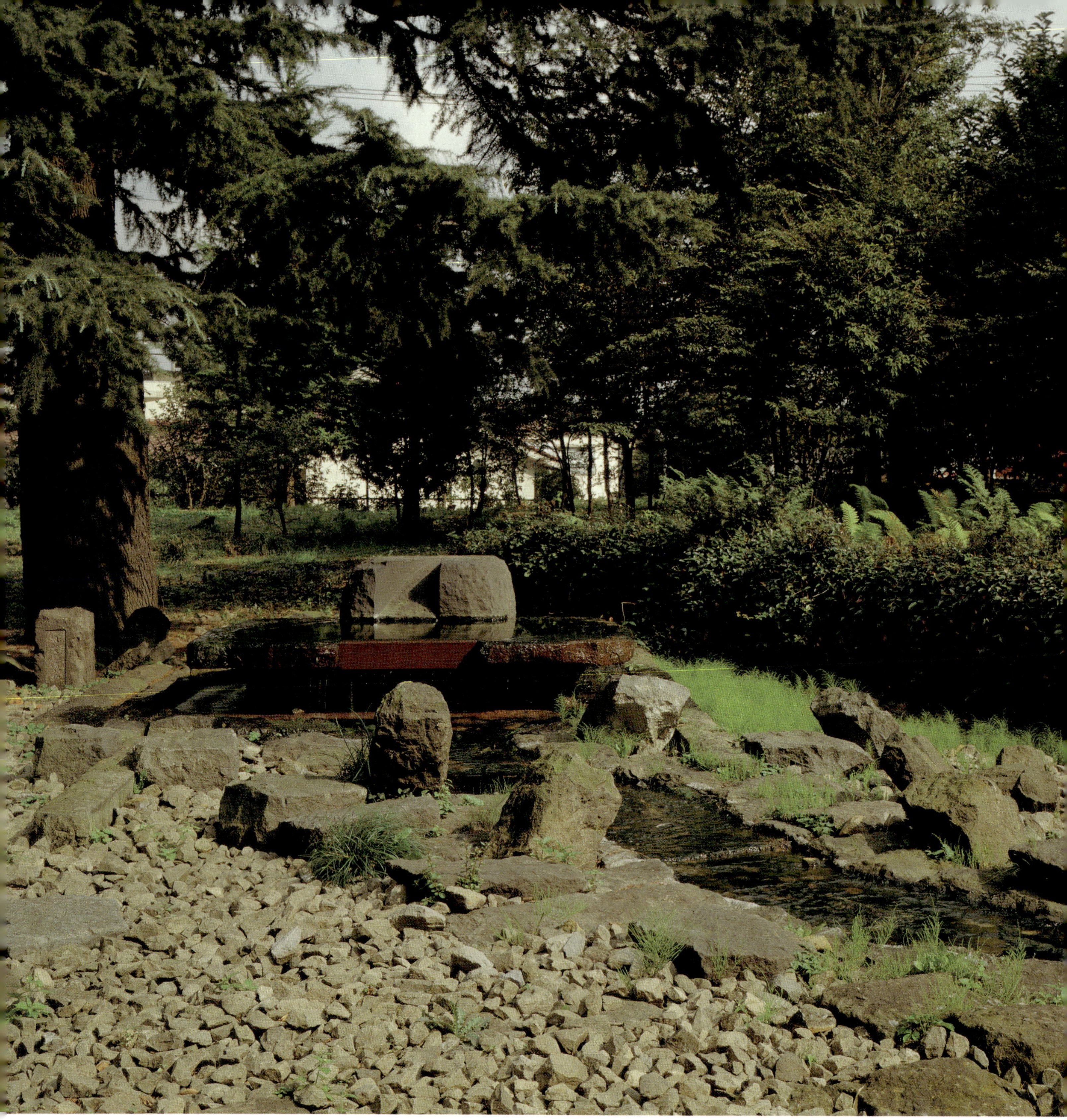

'Spring Water of Hino'

Bringing an object into a pre-made space means that the object would not exist independently in the space, but it must relate to the other objects and being conditioned, therefore, careful consideration was needed for its size, shape and position.

Besides, as the object was to be a commemorative present from all the executives and employees to celebrate their 40th anniversary at the end of the ten year period. Thus how to satisfy and rationalize the object's existence in the space became an important matter.

After thorough consideration of these two matters, making up the composition of the conceptual space its 'Forest of Hinodai' which was decision making at an executive committee meeting, blending it in to the surrounding landscapes, passing that consciousness on to later generations with the hope. And it became the size, shape and position seen at present.

As Hino City used to be a major supply of abundant spring water in Tokyo, the spring water was given shape, and its trail was reproduced and preserved in order to express the stability and prosperity of the company by expressing its abundance. On the other hand, the four concentric circles curved in the shallow water basin express each ten-year period since the establishment of the business while the last, fifth, circle is for the untrodden path, and this part is polished. So, what sort of conversation will be brought about by the V-shaped cut in the profile of the short stone pillar that holds the basin to viewers.

Shallow spring water is flowing into the moat. The clean water flows smoothly on the surface of the stone with no clearly defined water's edge. It is a short flow, but with time it will touch people's hearts. This too is the 'Spring Water of Hino'.

The 'Spring Water of Hino" will eternally convey its raison d'être to the people. It is a thing of value. The 'heart' is our treasure.
September 1982

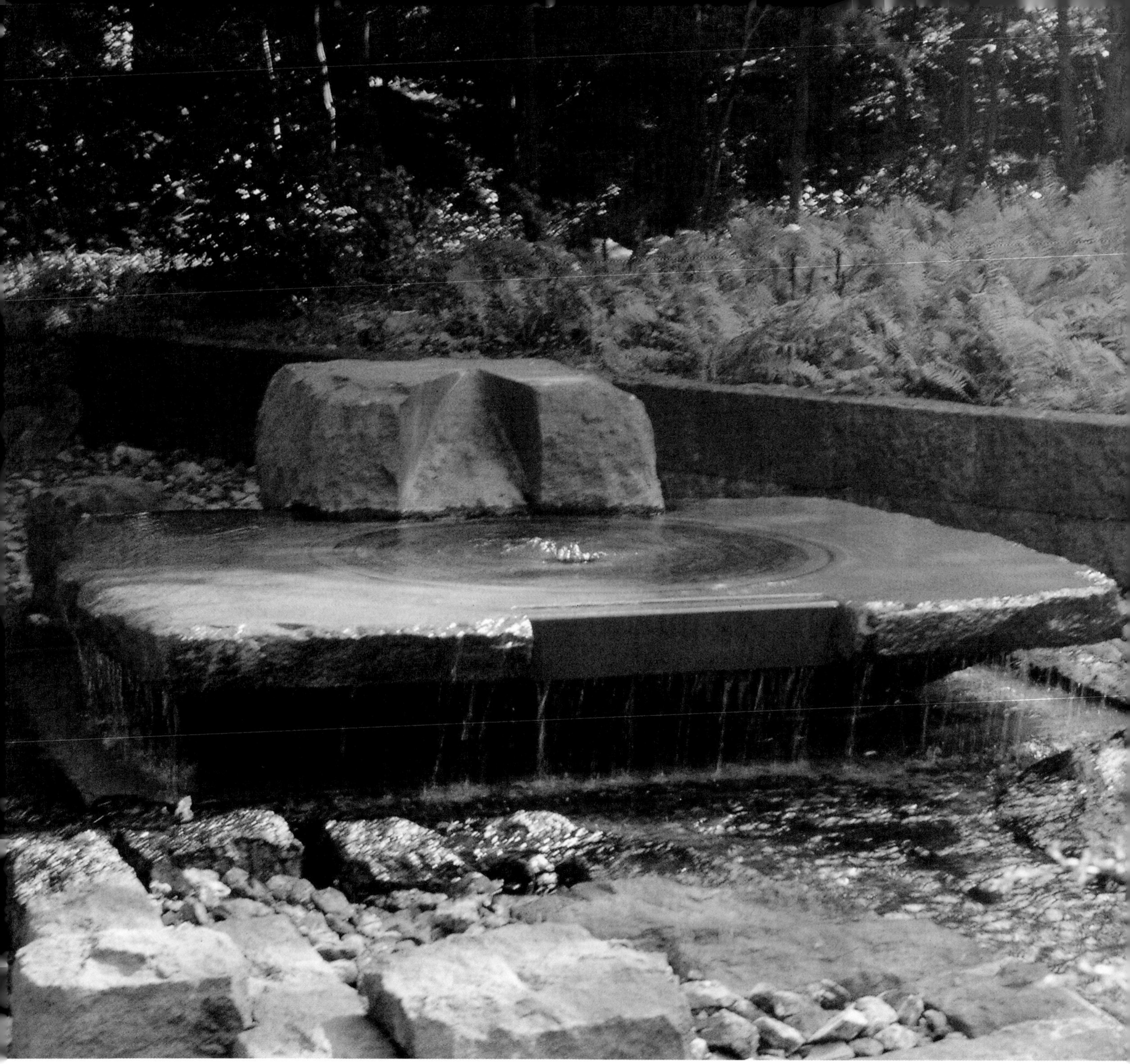

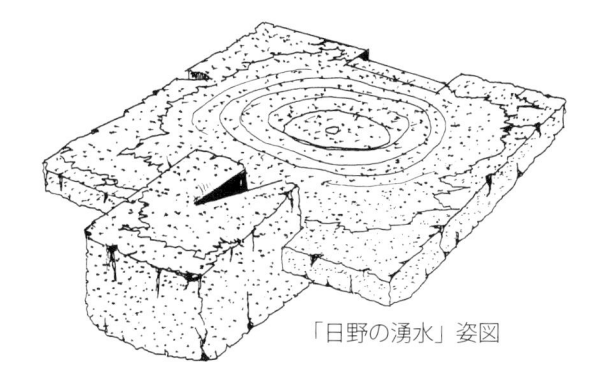

「日野の湧水」姿図

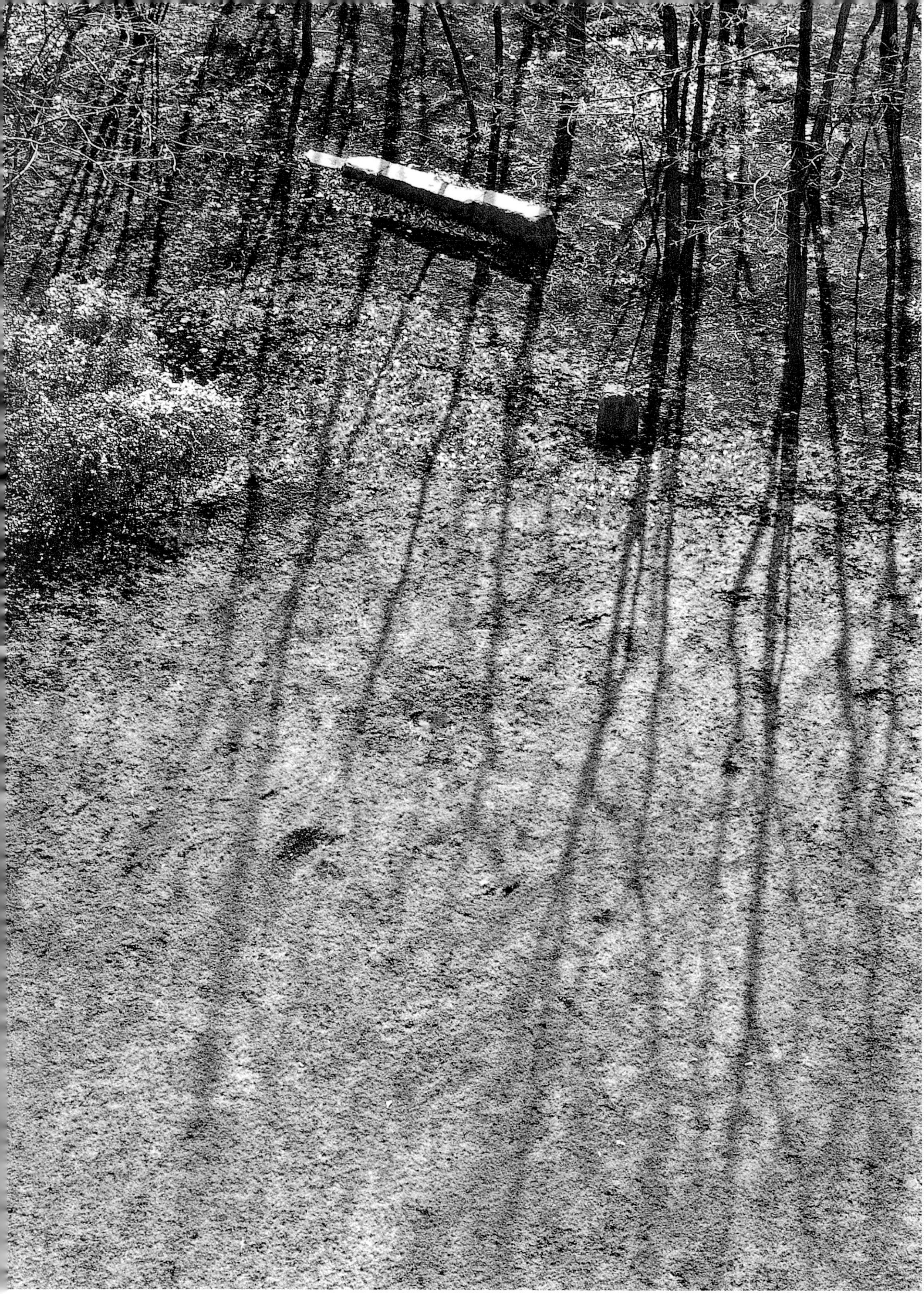

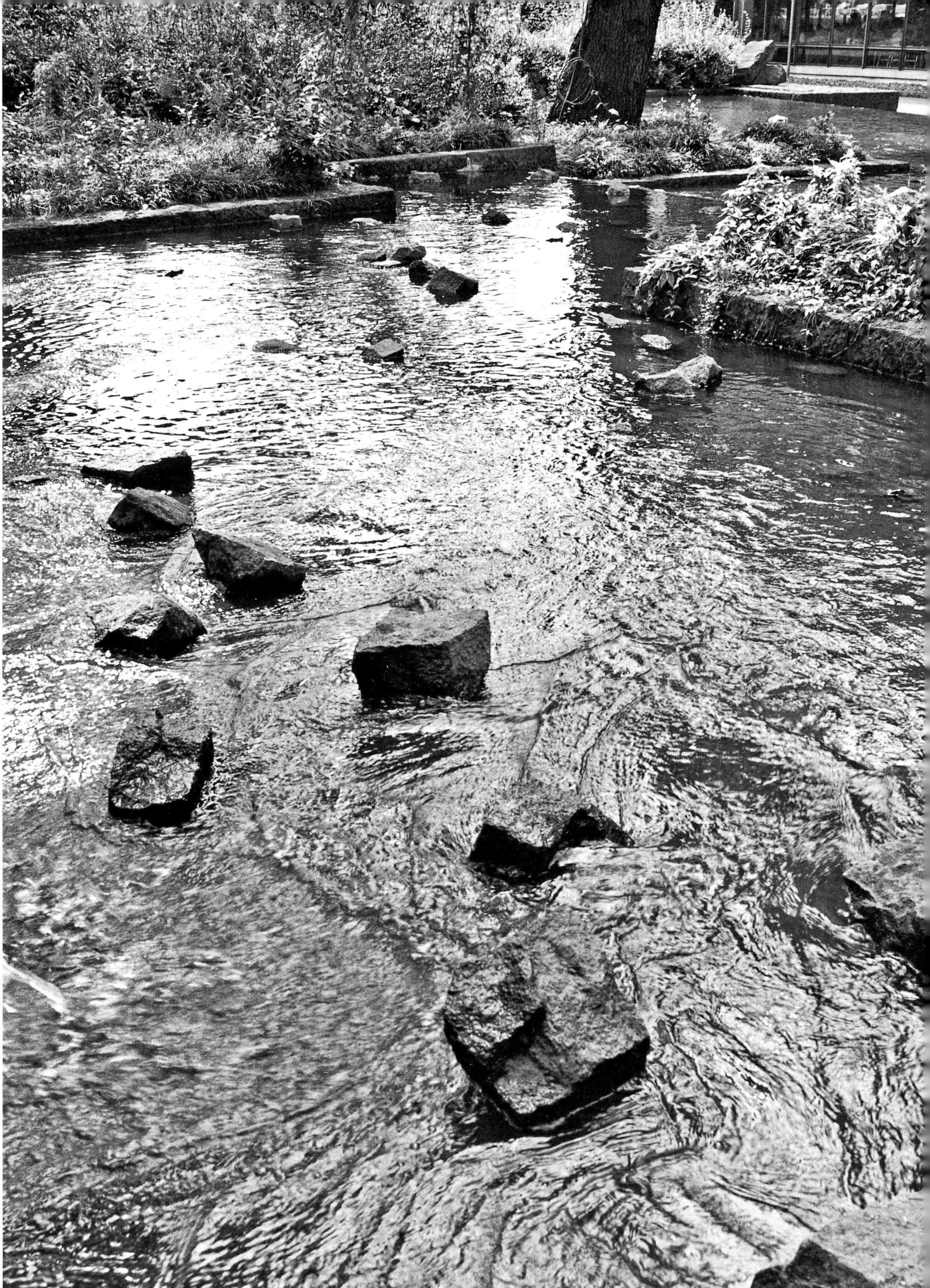

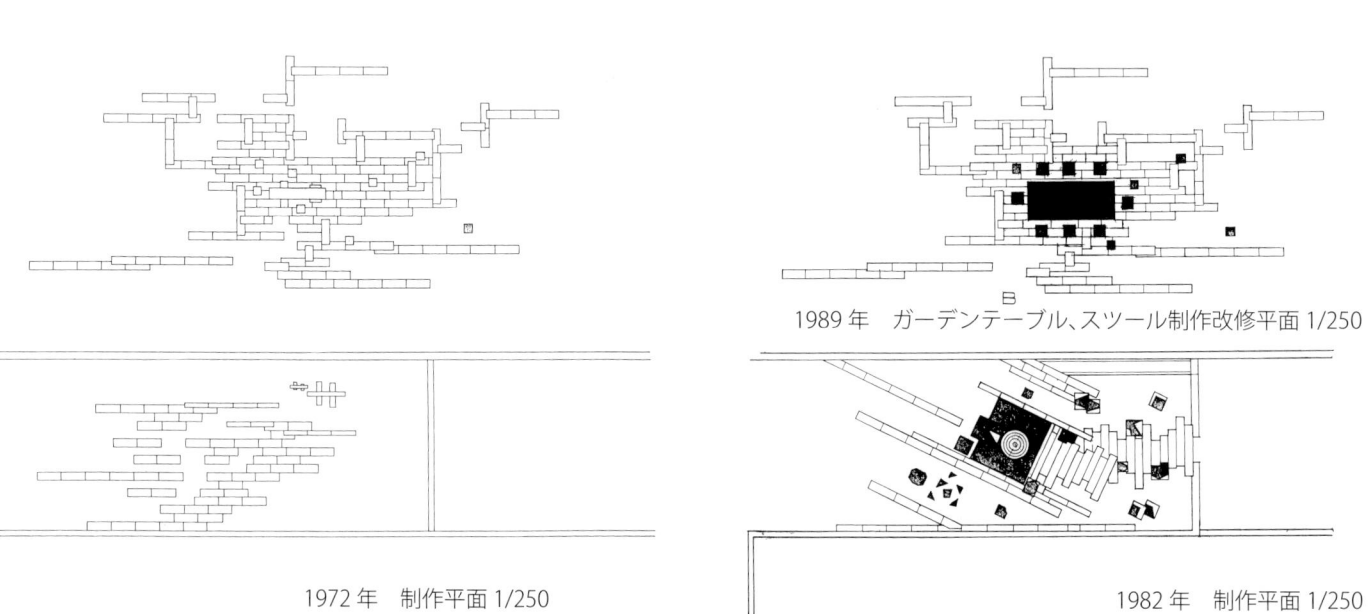

1989年　ガーデンテーブル、スツール制作改修平面 1/250

1972年　制作平面 1/250

1982年　制作平面 1/250

1972 年
用石の種類
石　積　　　　　　　　赤城眞石（間知石）、白河石（黒目）
ボーダー石、縁石　　　　白河石（黒目）、芦野石（黒目）
碑　　　　　　　　　　黒みかげ石ラステンバーグ
自動潅水設備　　　　　　ポップアップノズル
水の様態　　　　　　　　流れ、滝、堀

1982 年
用石の種類追加
記念碑（日野の湧水）　　安山岩、新小松石

1989 年
用石の種類（改修と追加）
カスケード縁石　　　　　黒みかげ石ラステンバーグ
ガーデンテーブル、スツール　　　〃
堀内植栽框　　　　　　　安山岩、白河石（黒）
林内舗石　　　　　　　　花崗岩、稲田石
空気抜蓋　　　　　　　　花崗岩、眞壁石

※詳細図はディテール図面編の 168～170P 参照

1972
List of Stones
Stone walls: 'Akagima-ishi', Kenchi-rocking-wall. 'Shirakawa-ishi' (Kurome).
Edging stone, stone curb: 'Shirakawa-ishi' (Kurome). 'Ashino-ishi' (Kurome).
Stone sculpture: Black granite 'Rustenburg'.
Automatic irrigation system: Popup nozzle.
Water feature: Stream, waterfall and moat.

1982
Addition to the stone list
The commemorative stone sculpture, 'Spring Water of Hino':
Andesite, 'Shinkomatsu-ishi'.

1989
Amended list of stones
Edge of the cascade: Black granite 'Rustenburg'.
Garden table and stools: same as above.
Edging curb inside of the moat: andesites, 'Shirakawa-ishi' (Kurome).
Curbs in the woodland: granite, 'Inada-ishi'.
Lid for a ventilator: granite, Kenchi-rocking-wall.

※ details 168-170p

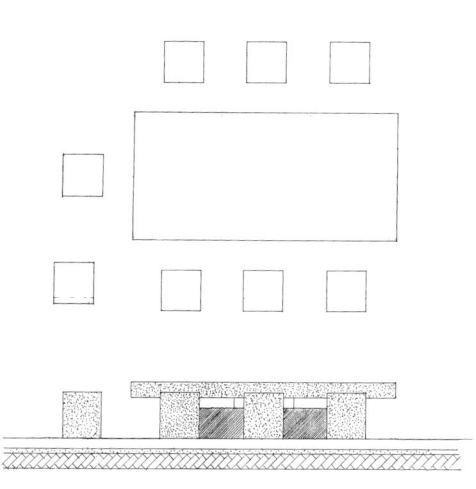

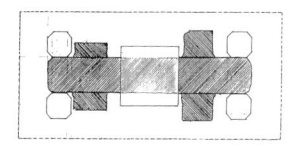

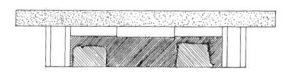

ガーデンテーブル、スツール N 立体 1/80

ホテルグランドパレス 外空間 1972
Hotel Grand Palace *'Gat-Kukan'* 1972

平面図 1/400

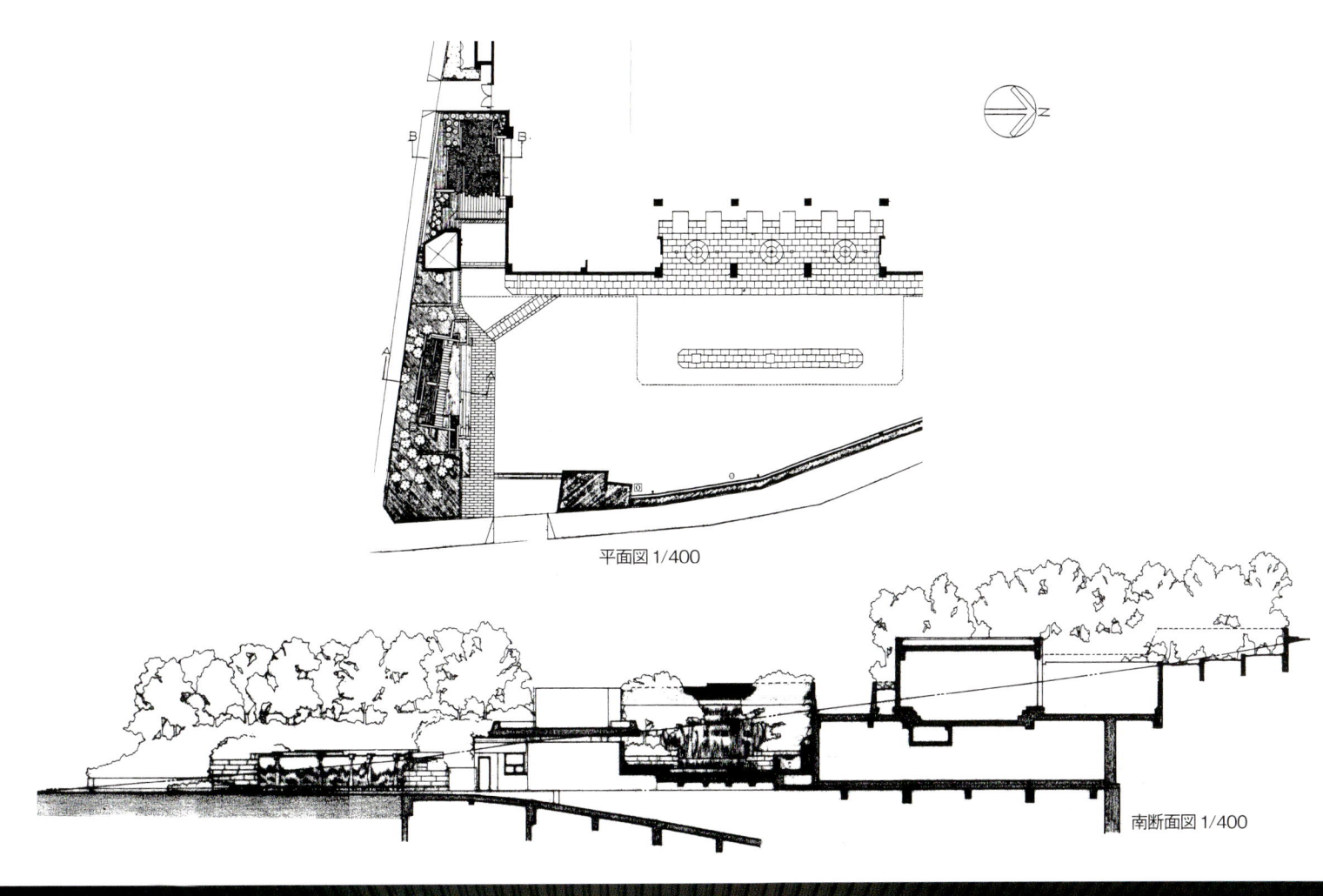

南断面図 1/400

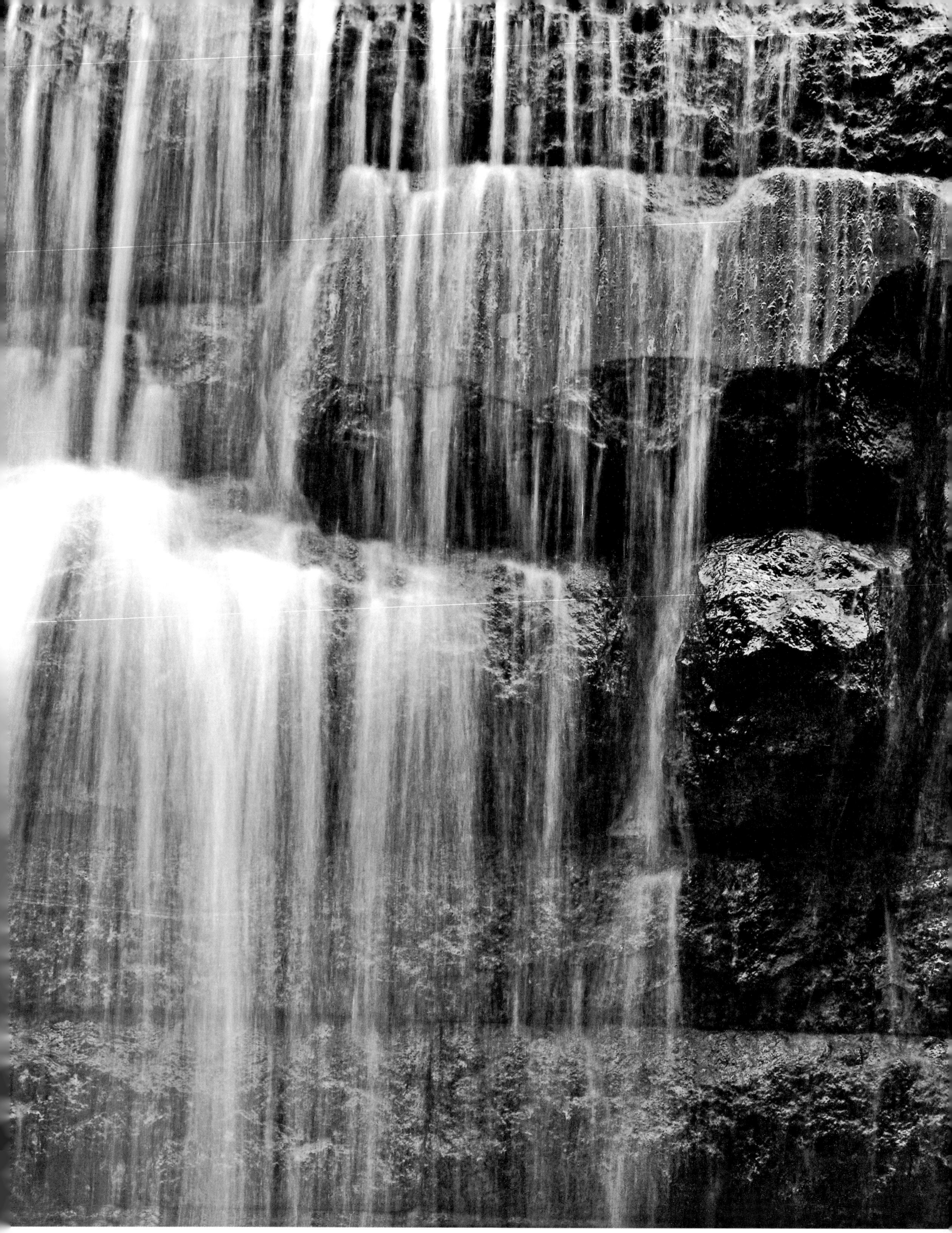

断面図 1/100

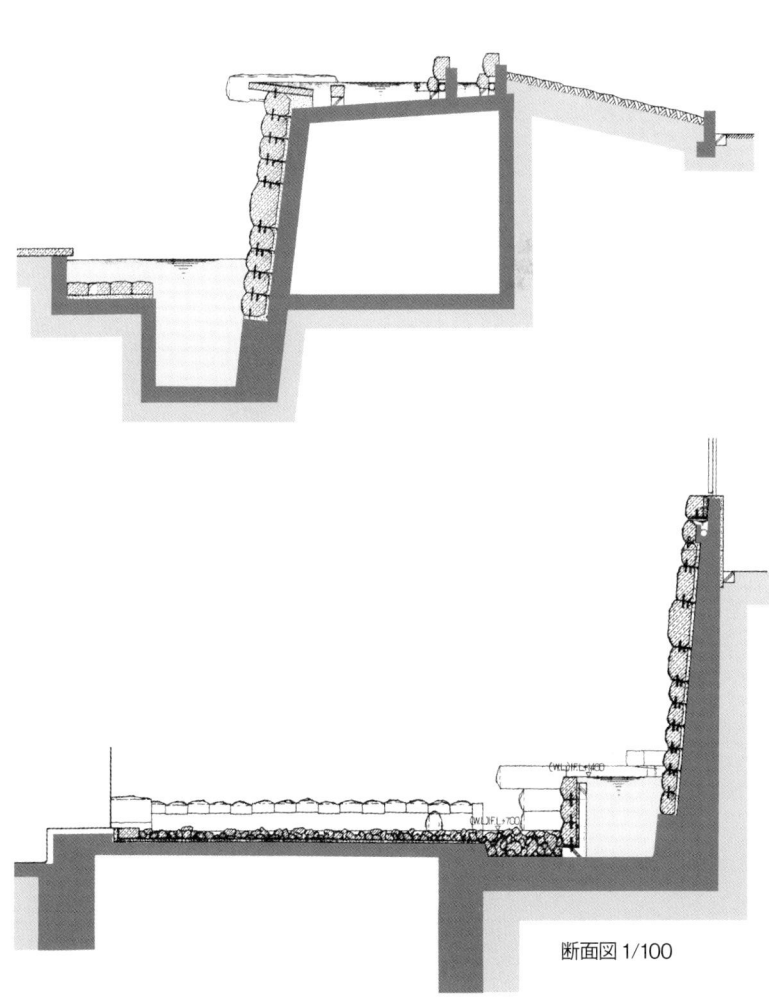

ホテルグランドパレス 外空間 1972

　この空間が依頼されたのは1971年早春の3月であった。

　このホテルの外部空間から、ホテルの機能を果たすための車路・駐車スペースなどを除くと、ホテル建物の東南側壁とフィリピン大使館からホテル前の目白通りに至る冬青木坂との間の図のような細長い異形空間が、外空間用地として残った。

　当時私は、京王プラザホテル、日野自動車工業本社の外空間の制作の最中であったこともあって、このような様態の空間に何を創り出そうか、考えをまとめるのに四苦八苦した。

　苦しんだ揚句、ホテルの立地する九段坂下の界隈の地に刻まれているのであろう、先人の生活の歴史を掘り起こすことを発想の手掛かりとすることに、考えをまとめた。

　考えあぐんだ末の神頼みとでもいうことであろうか。

　少し詳しく書くと、はじめに話したように何を創り出したらよいか悩んだ末、考えをまとめようとしばしば現場を訪れ、私の内部に伝わってくる何かを手掛かりにしようとした。

　九段の崖下のこのホテルの敷地の辺りに佇む度に、私の内部が古代へとイマジネーションをかき立てるのだ。これは、この地に宿っているであろう神のお告げであったのだろうか。

　九段下のこの辺りは、貝塚など古代人の生活の跡の発見・発掘の事実。

①九段会館の牛ケ淵貝塚に発見された弥生・土師遺跡
②現在、住宅・都市整備公団本社建設の折発見された、九段上貝塚など古代人の生活跡の発見・発掘

　他方、私の内部をしきりにまさぐる古代へのイマジネーションを重ね合わせ、弥生という農耕文化時代に思いを馳せ、発掘された様態でこれを神と結んだ。

　こうすることによって、この崖下では農耕と潮干狩によって安定した生活が営まれていたことになる。

　『外空間』（深谷光軌著・村井修撮影、1975年、誠文堂新光社刊）の冒頭に、神代雄一郎氏が「ヒューマンスペースの論理」という表題で一文を掲載しているが、その中で氏は、あたかも私の内部をまさぐっているかのような筆致で、この空間を語っている。

　こう見破られたのでは、さすがの私も「いや早や参った参った」と言うところである。

1972年
用石の種類
石　積　　　　　　　新小松石切石、芦野石切石
化粧土留ボーダー石　芦野石切石
化粧土留　　　　　　鉄筋コンクリート打放しツツキ仕上
水の様態　　　　　　滝、流れ
自動潅水装置　　　　スプレーノズル式

Hotel Grand Palace '*Gai-Kukan*' 1972

It was early spring, March 1971, when I commissioned this project.

A narrow irregularly shaped site shown in the drawing was left after cut off the followings from the external space of the hotel: the service road and parking space of the hotel, the east wall of the hotel building, and a section of Mochinokizaka from the Philippine embassy to the Mejiro-dori which runs in front of the hotel.

As I was in the middle of building '*Gai-Kukan*' for Keio Plaza Hotel and Hino Motors, Main Office, consequently I had a hard time to come up with good ideas for such a difficult site.

After careful and irksome considerations, I decided to observe the historical background of the hotel site, around Kudanzaka-shita, re-discover the predecessor's life history that may have remained under the ground for clue.

In a way it was like praying to god for help after being at loss what to think.

To be more specific, as I mentioned at the beginning, after going through a difficult search for good ideas, I use to gone down to the site and tried to collect clues.

Every time I stood in the area near the hotel under the cliff of Kudan, my mind would stir up the imagination about the ancient times. Could this been a divine message from the tutelary deity?

As it turns out, a number of remains of ancient people's have been found in this area, such as shell mound.

(1)Remains of Yayoi and Haze were found in the Ushigafuchi shell mound in Kudan Kaikan.
(2)Remains ofancient people, like the shell mound at Kudan-ue, were found at the time of construction of the City Housing and Urban Development Corporation's main office.

Then I overlapped the images of ancient times that was a spark of inspiration, and feeling for the Yayoi era, known to be a farming culture, and tied them with a god just as it excavated.

In this way, I could tell that stable life of farming and collecting shellfishes was here, under the cliff.

In the prologue of a book "'*Gai-Kukan*'" (author:Koki Fukaya, photographer:Osamu Murai, 1975, published by Seibundo Shinkosha), Yuichiro Kojiro wrote a few sentences abour the space under the title of "Logic of human space," in a clairvoyant manner.

Well, I could only admire his power of observation, been seen through like that.

1972
List of stones
Stone walls: Shinkomatsu-ishi. Ashino-ishi, cut.
Ornamental edging stone: Ashino-ishi, cut.
Ornamental retaining edge: Cast concrete with steel bars with finished prodding.
Water features: Waterfall, stream.
Automatic irrigation system: Spray nozzle.

小西酒造東京支店「坪庭」外空間 1974
Konishi Brewing Company Tokyo Branch Office 'Tsuboniwa' *'Gai-Kukan'* 1974

「坪庭」という名称だが、この空間が、毎日新聞社発行の『続坪庭』という図書に「小西酒造東京支店の坪庭」と名付けて掲載されているので、ここでもそう呼ばせて戴くことにした。

　小西酒造東京支店は、東京都中央区日本橋茅場町一丁目1番地の高速道路都心環状線を東を背にして建つ、伊丹市の清酒「白雪」で名のある酒舗の東京支店である。この「坪庭」に至る界隈の眺めは猥雑そのもので、その上「坪庭」は、背後に高架の高速道路都心環状線が架かるという、凡そ酒舗の「坪庭」の環境として救いようのない背景を持っている。

　このような空間に何を生み出そうかといろいろ検討の末、視的緊張感を生ませるものの存在を視野のいずれに設けるかがポイントになろうと、考えをまとめた。

　「坪庭」は写真のように掘り起こされた様態で、神と酒の泉を表象する単体（素材：新小松石）を置き、これを泉として水を湧かせ流し、訪れる人びとの心に視的緊張感を生ませ、遠景として目に入る高速道路都心環状線を意識から外し、他方、大都市の猥雑さ喧騒の中に在ることを一時忘れ、静かなひとときを供しようといった意図で制作したものである。因みに、この「坪庭」を正面に見る内空間は見本展示室と店舗を兼ねた空間で、側面に見る空間はアプローチでありエントランスホールである。

　この空間は何時頃壊されたのであろうか現在はその面影はなくなった。淋しいことである。

1974 年
用石の種類
単　体　　新小松石
敷　石　　芦野石切石
水の様態　滝、流れ

61

平面図 1/200

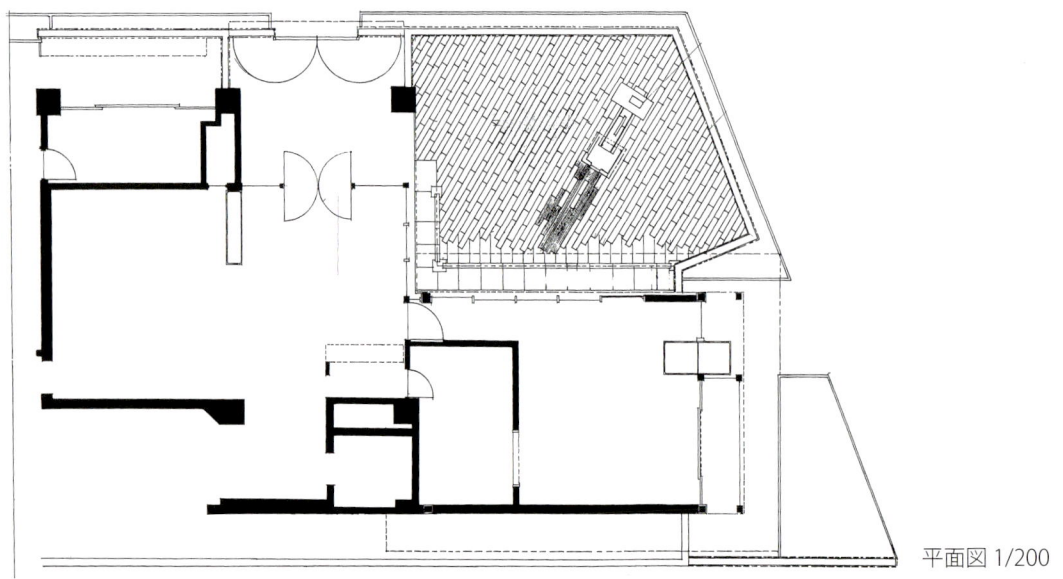

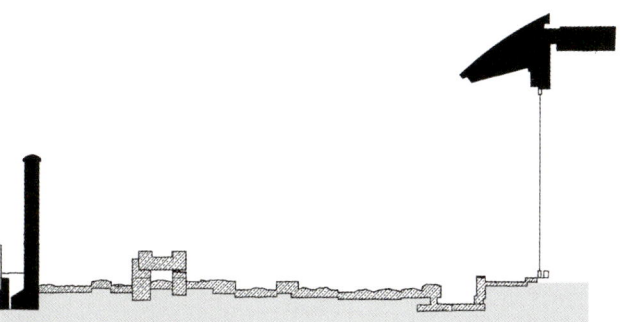

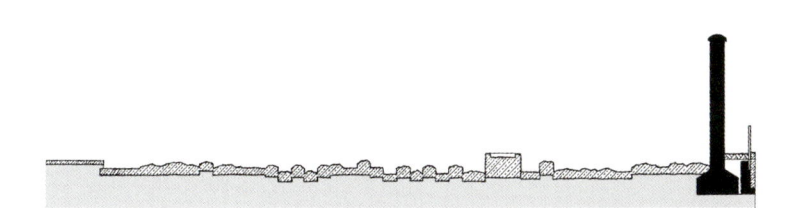

Konishi Brewing Company Tokyo Branch Office
'Tsuboniwa' *'Gai-Kukan'* **1974**

As this space was introduced as 'The Tsuboniwa (small courtyard) of the Konishi Brewing Company Tokyo Branch Office' in the book, "The sequel to the Tsuboniwa" by Mainichi Shinbun publishers, I will call it so here, too.

The site is located between the Konishi Brewing Company Tokyo Branch Office, located at 1-1 Kayaba-cho, Nihonbashi, Chuo-ku, Tokyo and an elevated highway, the metropolitan ring road to the east. The company produces a well-known Sake (Shirayuki – White Snow) and its origin is in Itami city, Hyogo prefecture. All in all it is a hopelessly terrible place and totally lacking in the elements required for a 'Tsuboniwa' of beauty, what with the elevated metropolitan loop bridges over the 'Tsuboniwa' in the background.

After thinking over various ideas, I decided that the thing to do was draw the attention of the viewer away from the surrounding ugliness, and the difficult thing to decide was where to locate this element.

The 'Tsuboniwa' is seen as if dug out, as shown in the picture, and a stone was put in place (material: Shinkomatsu-ishi) symbolizing God and a well of Sake, letting the water flow like a spring, thus removing the viewer's attention away from the highway, the metropolitan ring road at the edges of visibility. At the same time, this was designed to offer serenity in the midst of the noisy and gruesome mega polis. For comparison, the interior space that looks out to this 'Tsuboniwa' at the front functions as a sample gallery and shop, and the space at the side is the approach and entrance hall.

I wonder when this space was razed. There is no atmosphere left there now. It's really quite a pity.

1974
List of stones
Single stone: Shinkomatsu-ishi.
Stone paving: Ashino-ishi, cut.
Water features: Waterfall, stream.

断面図 1/100

東京銀行青山寮 外空間 1975
The Bank of Tokyo Aoyama Retreat 'Gai-Kukan' 1975

この外空間の写真に見られる雰囲気から、1976年に完成した和倉温泉「かがや」の別館「サンかがや」の外空間の雰囲気が彷彿とさせられるであろう。

　実は、この空間の依頼の1年半程前から「サンかがや」の外空間の計画が進み、制作模型を素に様々な検討が行われていた。その最中にこの外空間の仕事が依頼され、短い時の流れの中で計画と制作が進められることになった。そういうことになったので、「サンかがや」の空間の部分模型を用い、構成の検討などを行ったからであろう。

　この青山寮の建設地の旧地は、王室であった李王根家の屋敷跡といった由緒ある地で、築地の塀に昔時の王室を示す四條の線が刻まれた、印象深い雰囲気が遺されていた。

　依頼されたのは、敷地の内部に取り込まれた2つの方形の空間で、各々の面積が拡大か、あるいは狭少ならば非常に扱い易いが、正方形に近いこの空間の規模は、辺りの眺めを考慮に入れると、視覚的に様々な検討と構成と展開の検討にかなりの時間を要するので、非常に難しいもののように思われた。その上、失敗は許されないから様々な神経が遣われた。

　先に話したように、「サンかがや」の部分模型を使って構成の検討を行い、掘り起こされた様態で、神の存在したかつての日本の自然と、王家の屋敷の遺構を思わせる石の構築とで構成した。落葉樹の木々（雑木林の木々）が「自然な自然」で、石の遺構を思わせる構築が「意図された逞しい自然」である。

　通用門からエントランスホールに至る樹木の根方に、石を刻んで創られた大きな水鏡がある。

　水鏡は、訪れる人びとを、梢を通して映し出される雲の動きと青空と、都市生活の中で体験することが失われてしまった「自然な自然」の眺めに誘い込む。

　この空間の構成と展開が無理なくおさまっているのは、共に制作した石工頭の宮本貞八君と私との息が全く合っていたからであろう。

　京王プラザホテルの4号街路外空間とは、別の迫力を生んでいるように思えてならない。

　敷地に取り込まれた空間にしておくのは勿体ないと、常々考えている。

　この外空間の定かな設計図はない。私自身が自ら宮本貞八君と現場制作を行い、2人のハプニング制作で出来上がったからであろう。

　2人はこの仕事を終え、和倉温泉「サンかがや」の制作に向かった。

1975年
用石の種類
単　体　　　　　　　新小松石
石　積　　　　　　　芦野石切石、芦野石雑石
ボーダー石、縁石　　芦野石切石
敷　石　　　　　　　芦野石枝石
化粧砂利　　　　　　鞍馬石五郎太

66

The Bank of Tokyo Aoyama Retreat *'Gai-Kukan'* 1975

Looking at the photograph of this *'Gai-Kukan'*, one would remember *'Gai-Kukan'* of the annex "Sun Kagaya" of the hotel "Kagaya" of Wakura hot spring resort built in 1976.

As a matter of fact, the planning of the exterior space with a model for "Sun Kagaya" had been going on since about one and a half years ago, when I was asked to produce this exterior space.
This order came in the midst of the production process of the other space and therefore I had to go on with the planning and production in a short period of time. Under the circumstances, a part of the model of "Sun Kagaya" was used in contemplating the composition of the other space.

The site of construction of the Aoyama House was a historical place where was a royal family of Ri Okon's mansion, and on the earthen wall surrounding it were carved the four lines signifying the status of the royal family in old days, giving an impressive atmosphere to the place. On conditional, the designing of order was two nearly square spaces entirely. It would be easier if it was more larger or smaller, it seems to be very difficult and time consuming to plan, due to the considerations of the visual aspect, the composition and development of the space, under the circumstances, it should be pay attention severely, caused, can not mistake it was.As mentioned earlier, the model for the "Sun Kagaya" was utilized design that was based on the nature of old Japan in which god lived and a stone structure that like a reminds such the old residence of a royal family seemed be excavate. Deciduous trees and woods are the "natural nature" and the stone structure like a remains is the "intentionally bold nature".
Near root of a tree from the side gate to the entrance hall, there is placed a large of the water as mirror on a carved stone. that reflects the sky and invites the viewer to see the movement of the clouds and blue sky through the branches of the trees.
It would be catch a glimpse of the "natural nature" that are now lost from experience in the urban life.

The fact, settled fitting perfectly the space by the design and construction, it's due to the perfect matching with Sadahachi Miyamoto, the head stonemason who worked with me together the projects. I'm feeling it has another kind of atmosphere power as such the *'Gai-Kukan'* at Keio Plaza Hotel on the 4th street. I thought it is too excellent to be confined within a closed premise in this area.
Nothing exact drawing for this *'Gai-Kukan'*, but the result is superb caused something like a creative happening by two of us. For thanks to worked with Miyamoto in the field.
After finishing this work, we started the project of "Sun Kagaya" at Wakura Hot Spring.

1975
List of stones
Single stone: Shinkomatsu-ishi.
Stone walls: Ashino-ishi, cut. Ashino-ishi, fragments.
Border stone, Stone curb: Ashino-ishi, cut.
Stone paving: Ashino-ishi, branch stones.
Ornamental gravel: Kurama-ishi Gorota.

能登「サンかがや」外空間 1976
Sun Kagaya in Noto '*Gai-Kukan*' 1976

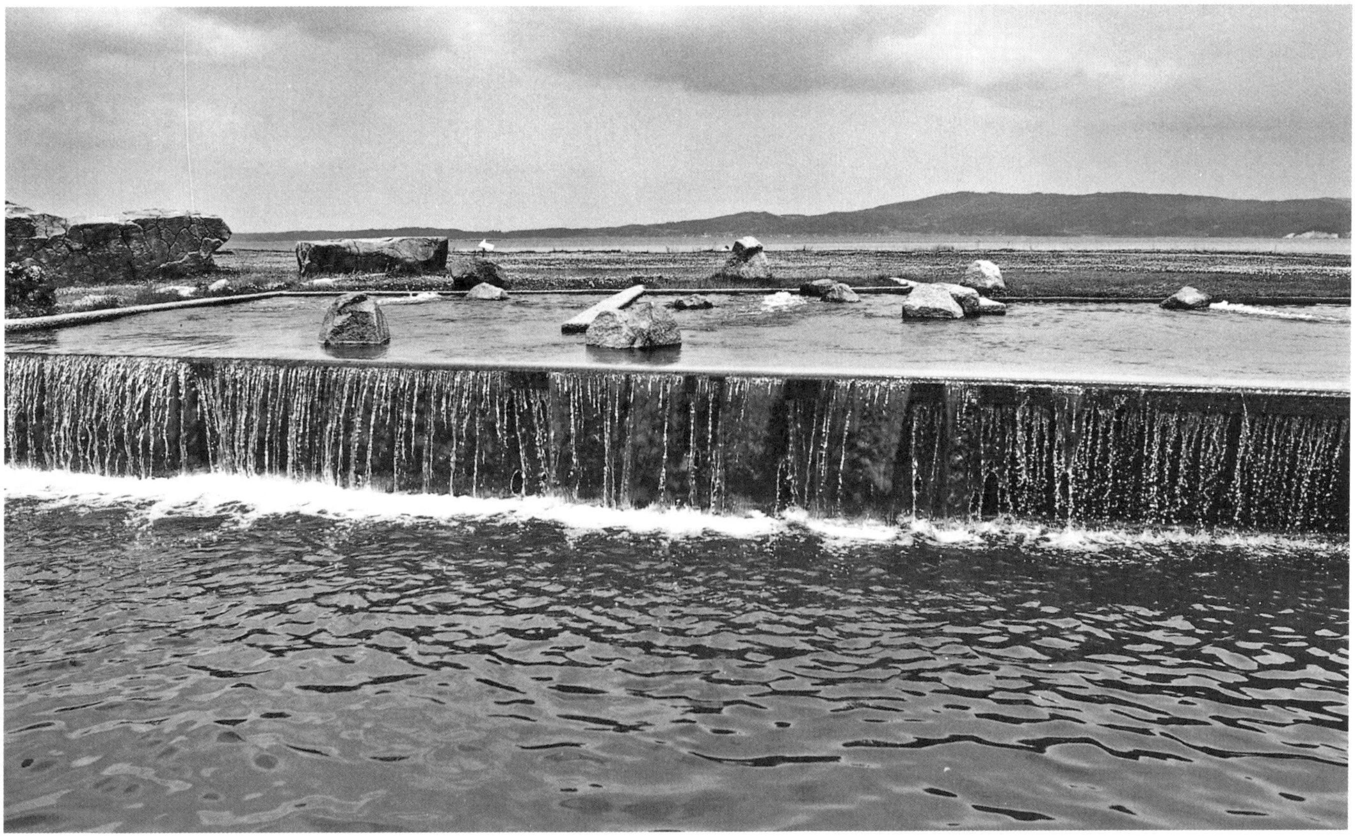

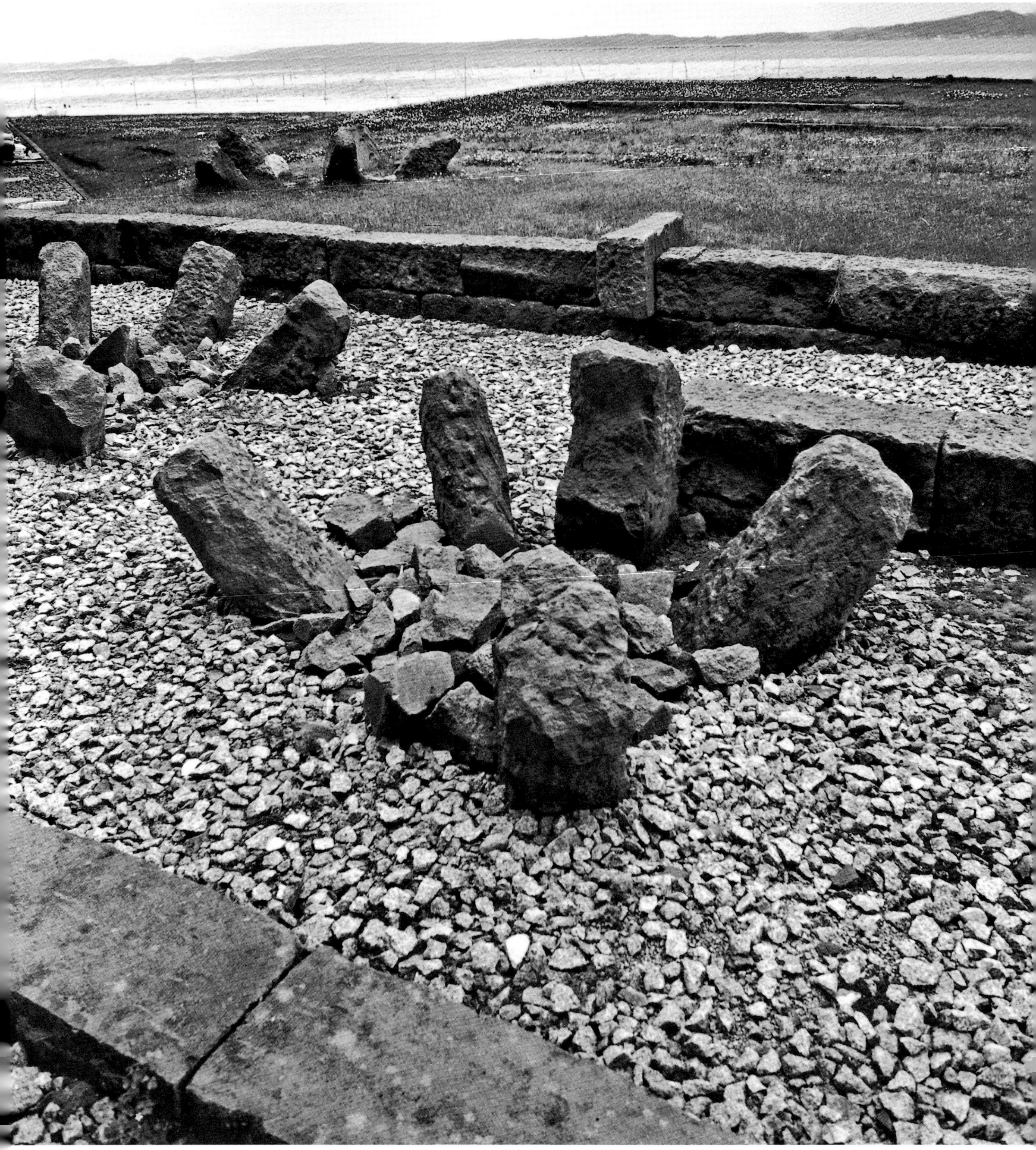

71

72

| 能登「サンかがや」外空間 1976 | Sun Kagaya in Noto 'Gai-Kukan' 1976 |

心で見る世界、心で創る世界など。心（精神）とは何かを失いつつあった難しい流れの中で、夢のような心（精神）の世界への憧憬。そしてそれへの道が拓かれるのは唯一抵抗の精神から生まれるのではなかろうかとした、純粋さあふれる内面の充実した時の流れであったような思いがしてならない。

「サンかがや」の外空間は、そのような心の移ろいの最中に生まれた空間である。

この空間は、当時能登地方に遺る（現代も遺っているであろうか）土地に宿る神々への祭祀を、私の内部が石に託して謳い上げた空間である。

空間は敷地の内部に取り込まれているがオープンスペースの中でも充分成り立つ。それだけ人びとの心に割り込める意味の深さを持っていると思う。

いつ写真を見ても、心和む空間である。

親しい人の噂によると大分壊されたという。身のまわりから人の手が創った痕跡が失われて往くのは淋しいものである。福島県二本松出身のガンコにして単純な精神（こころ）の持主の石工・宮本貞八君があの世で地団太踏んで怒っているだろう。生前の彼が最も情熱を傾けて創っただけに。

1976年
用石の種類
単　体　　　芦野石雑石
石　積　　　芦野石雑石、石動山みかげ自然石
堀（構築材）　鉄筋コンクリート打放しツツキ仕上
テラス舗床　　プレコン板
水の様態　　　滝、流れ、堀

Focusing creativity of our mind. At a time when we tend to forget such a mind (spirit) really is, it would be still pure and rich occasion, having adoration toward the dreamlike world of mind (spirit) and that can lead us the only spirit of resistance to such a world.
'Gai-Kukan' of "Sun Kagaya" created just at the time in such a feeling of my mind.
In those days, the people in Noto district used to offer prayers to the god of land.
Thus, this space was performed with my own spirit by the formed stones as very prayers.
This 'Gai-Kukan' is confined in a premise but it's just effective if it was in open space. It would be giving deep impression for the mind of people as such a profundity meaning.
This 'Gai-Kukan' photographs, what I taken, always gave me a peace of mind.

According to friend of mine, the space has been damaged almost. It is a sad that something made by my own hand or others is lost. Sadahachi Miyamoto, a stonemason, from Nihonmatsu, Fukushima prefecture, who was stubborn, had pure spirit (heart), must be stamping his feet with anger up in heaven, because This 'Gai-Kukan' had most bring in all the passion by himself.

1976
List of stones
Single stone: Ashino-ishi, fragments.
Stone walls: Ashino-ishi, fragments. Sekidozan, natural granite.
Moat(structural material): Cast concrete with steel bars finished with prodding.
Paved terrace floor: Pressed concrete plates.
Water features: Waterfall, stream and moat.

東銀栗田ビル（現 一ツ橋 SI ビル）外空間 1978
Togin Kurita Bldg. (the present is Hitotsubashi-SI-Bldg.) 'Gai-Kukan' 1978

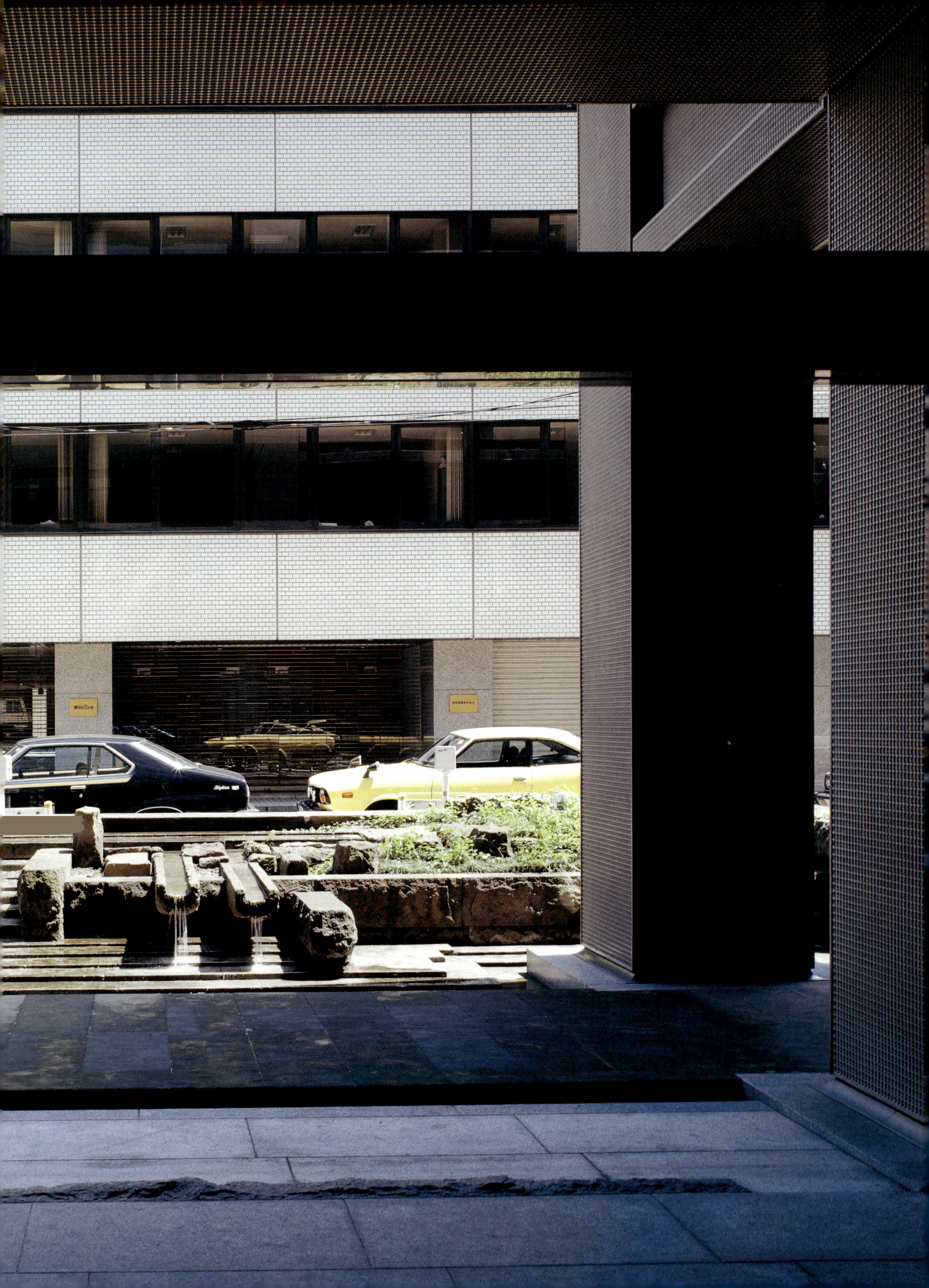

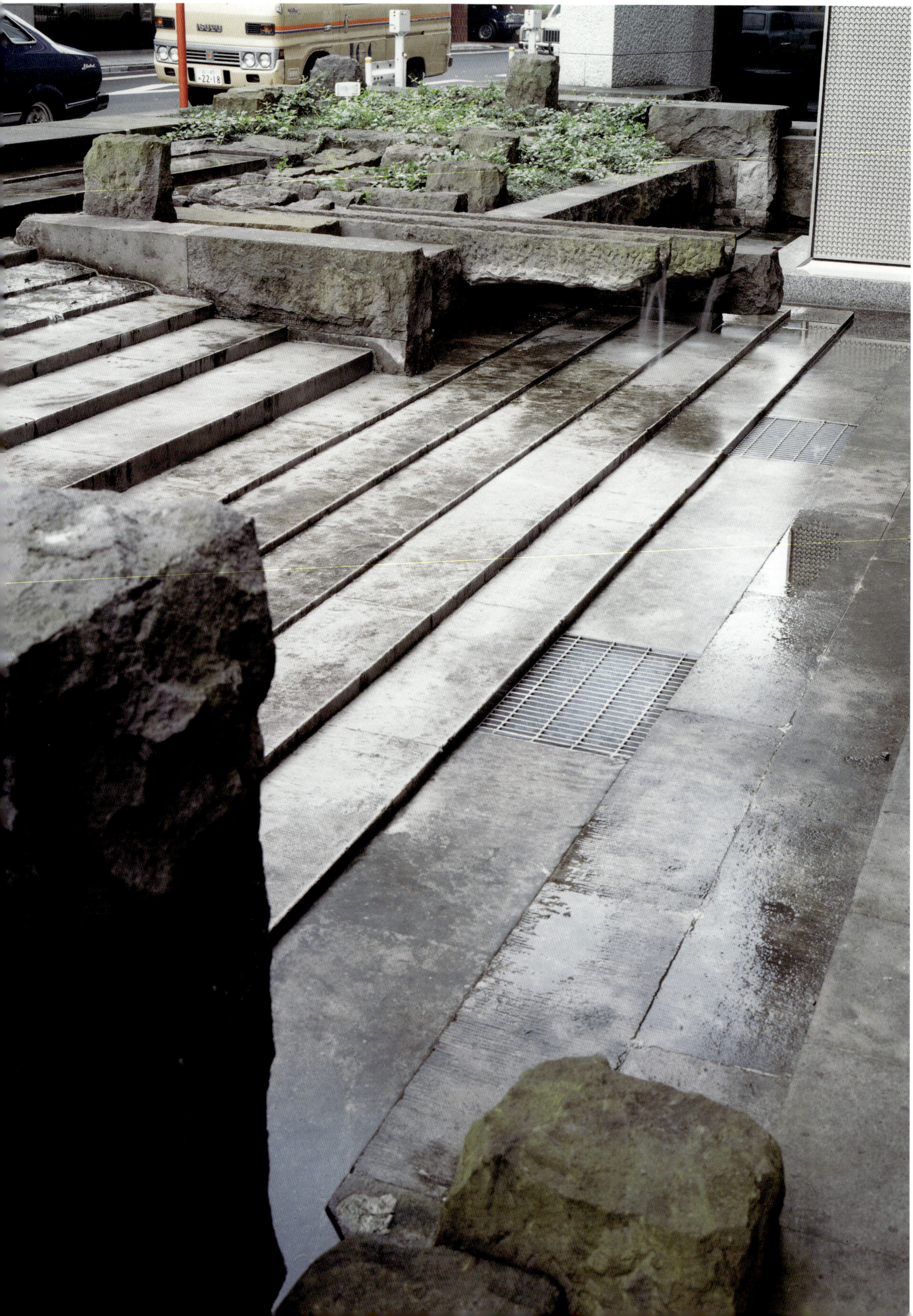

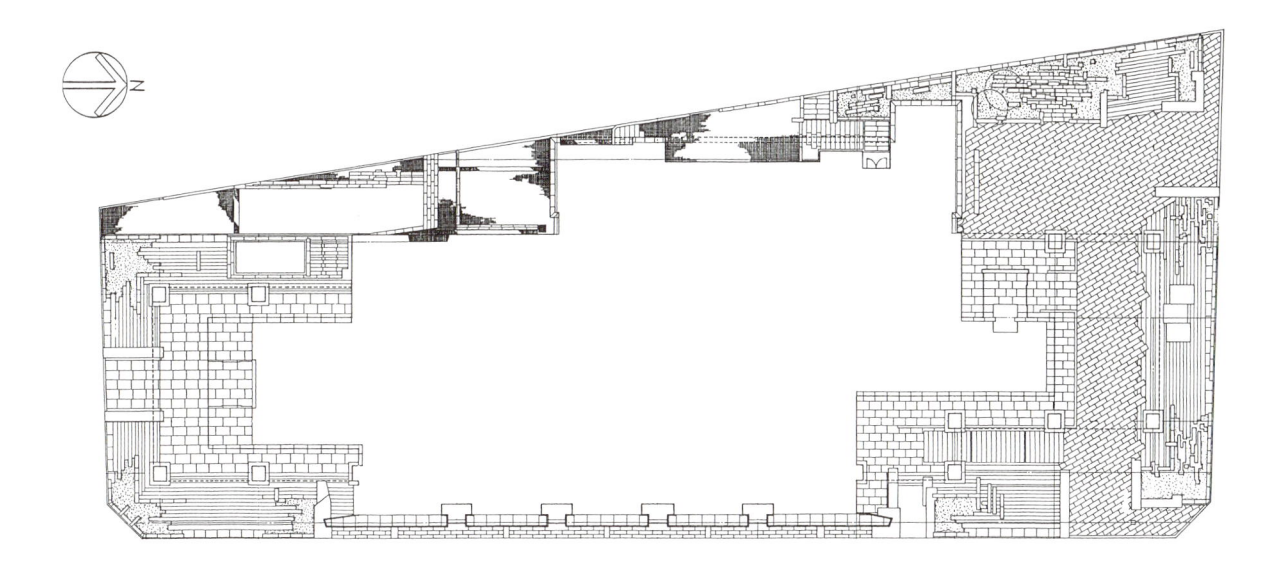

配置平面図 1/600

78

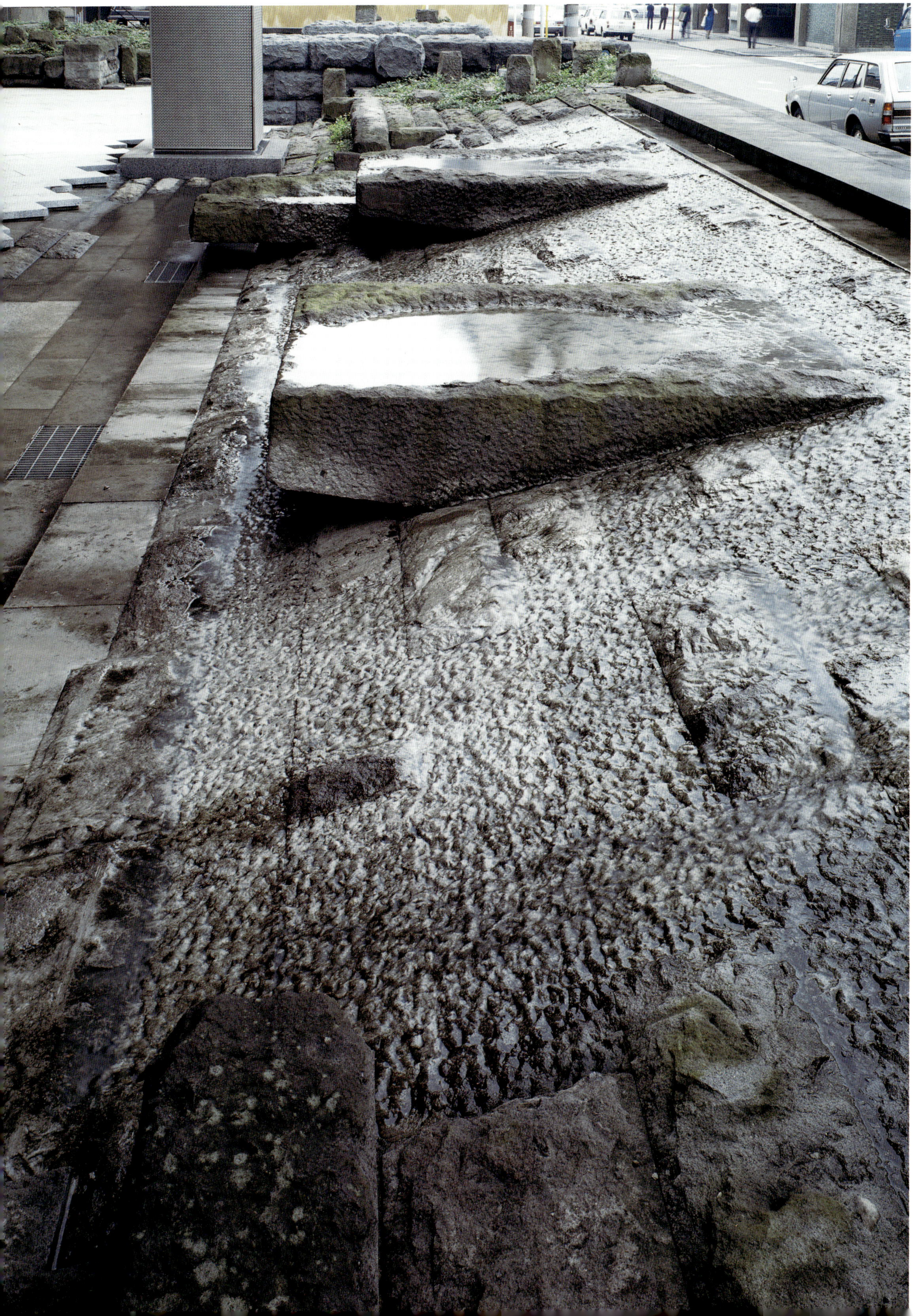

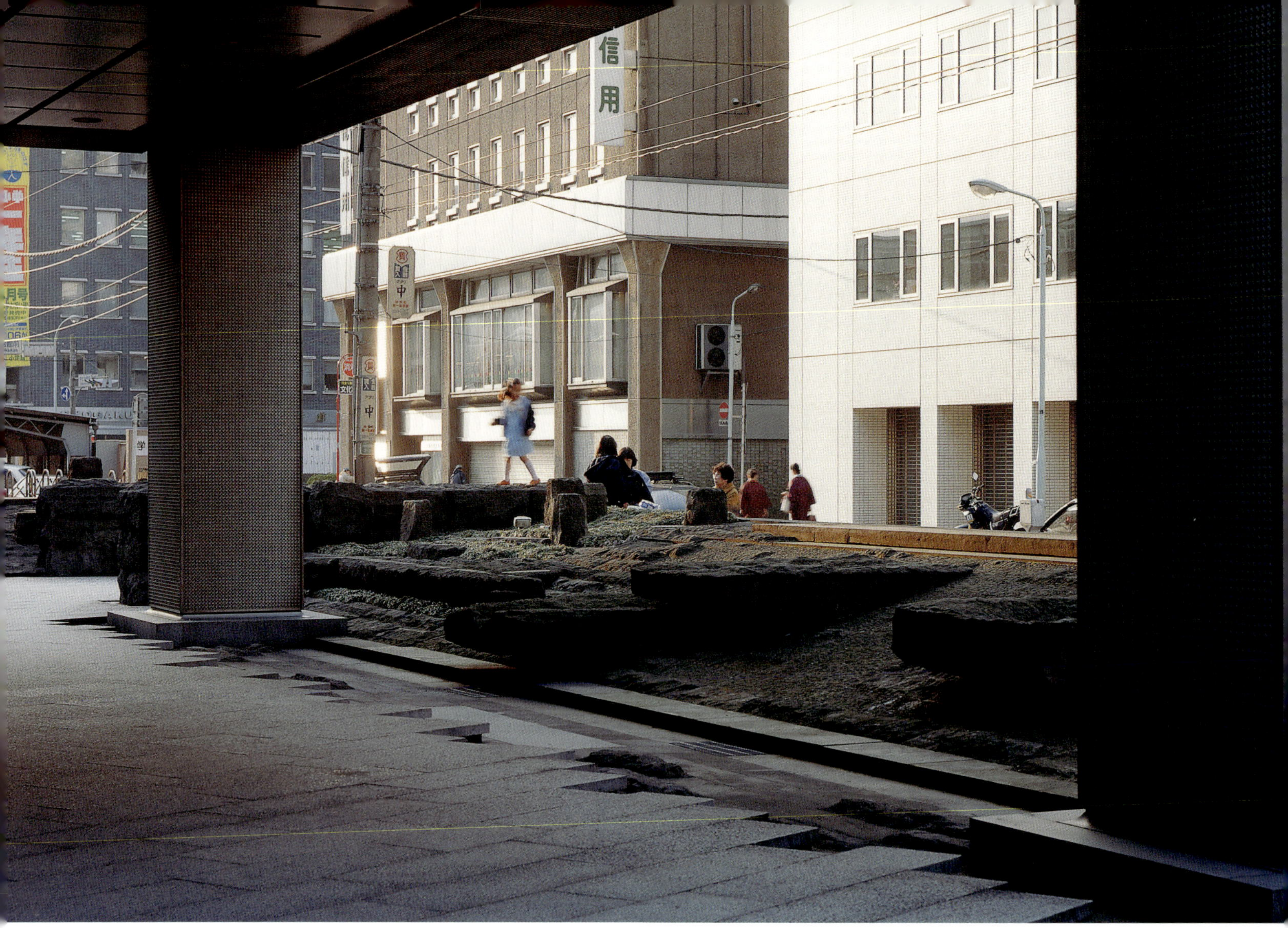
信用

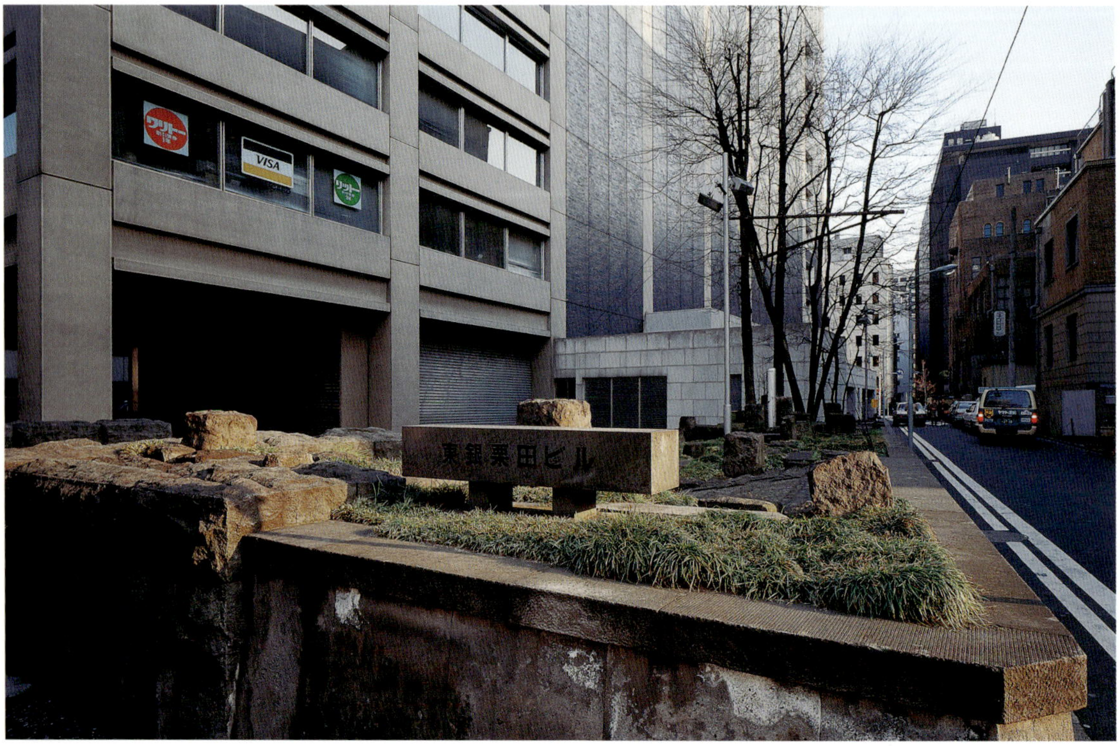
VISA
東銀栗田ビル

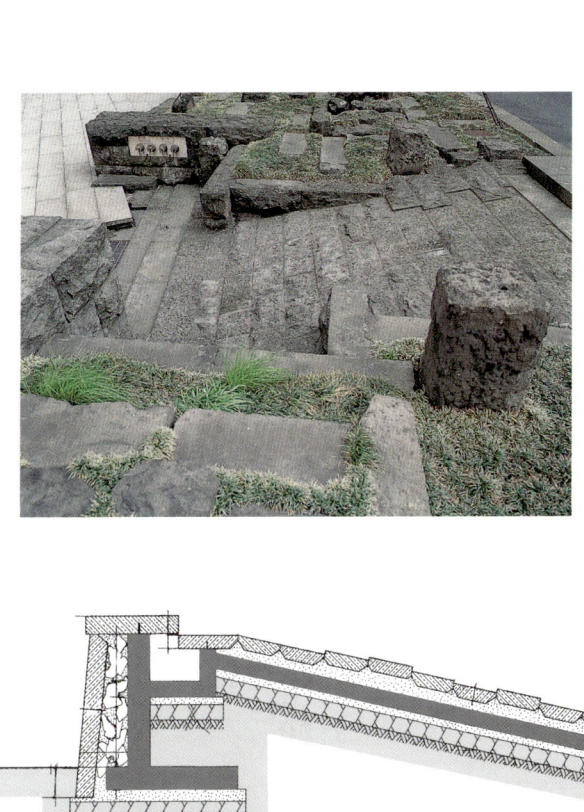

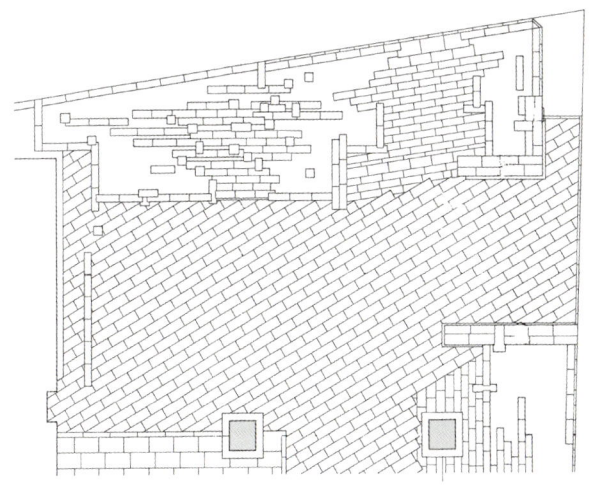

平面図 1/300

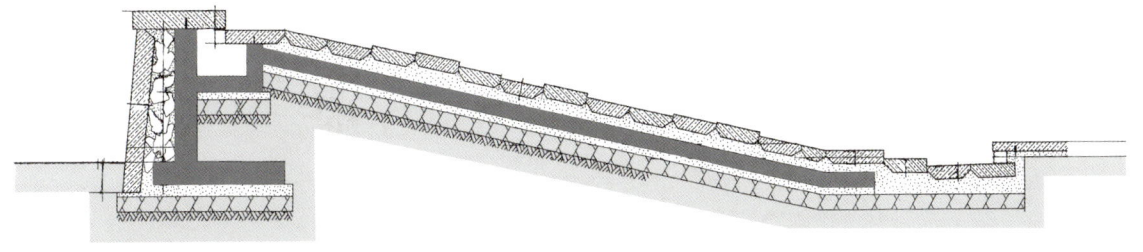

断面図 1/50

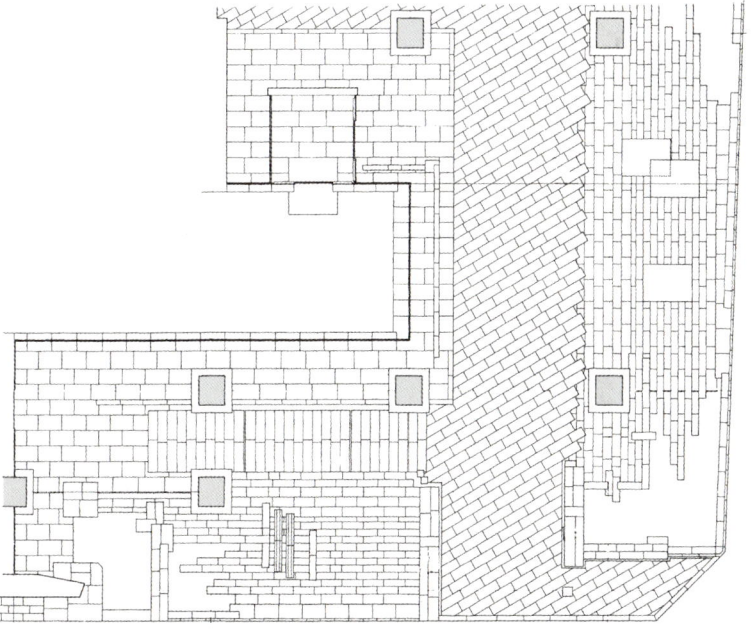

平面図 1/300

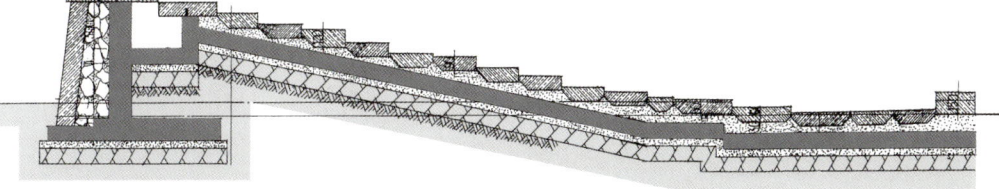

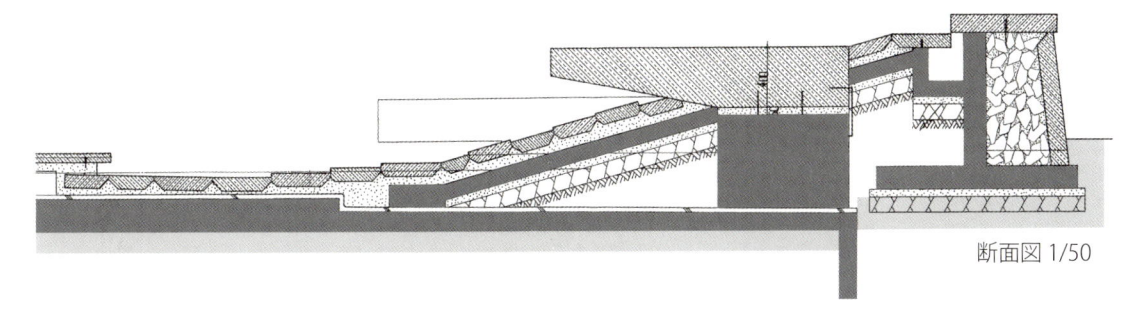

断面図 1/50

平面図 1/300

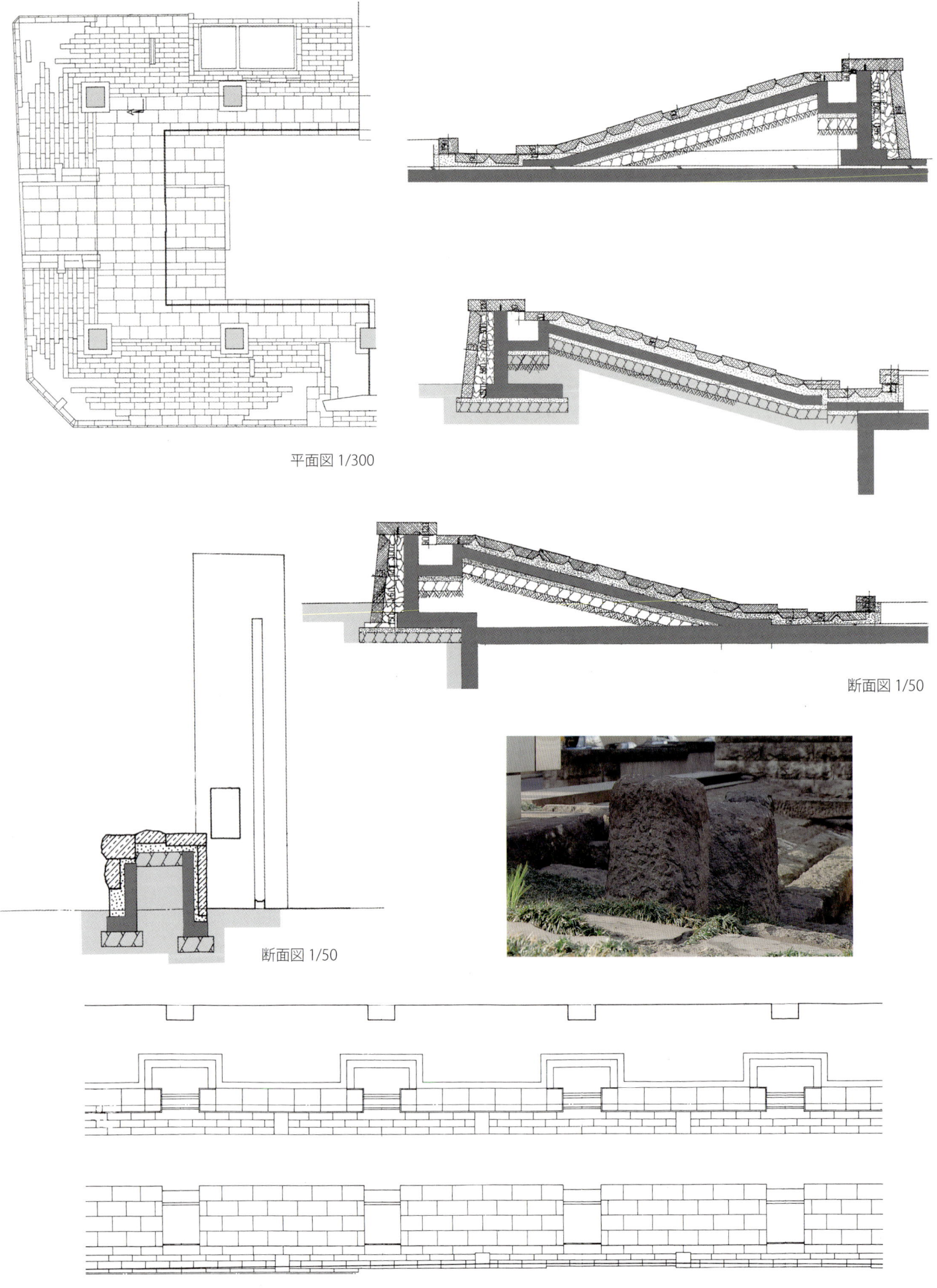

断面図 1/50

断面図 1/50

立面図 1/200

82

83

東銀栗田ビル（現一ツ橋 SI ビル）外空間 1978

「東銀栗田ビル（現一ツ橋 SI ビル）」は、図のような位置に立地する建物の東・西・南・北面それぞれが一方通行路に取り囲まれている敷地に、多少の残地を有して建つ、外壁がアルキャストの 10 階建てのビルで、1978 年 12 月完成した。

東銀栗田ビルの外空間は、再開発の進みつつあった神田一ツ橋の業務地区の公共空間に割り込んで、ビル敷地と建物を背景に道（公道）と一体化することに成功している。

ことに北側のビルエントランスは、附近に公共空間の乏しいことからピロティとして公共に開放しているので、エントランス内部からの眺めと公共空間としての扱いが一体化され上手に処理されている。

それぞれの空間に視的緊張感を生ませる点について充分検討しているので、その点でも学ぶ処が多いと思う。

このビルの建つ辺りは江戸時代火事が多く、現在の靖国通りから濠端まで、東は神田橋から小川町交差点、西は九段下までの小川町低地南部には、広大な防火地帯が造られていた。元禄期に五代将軍綱吉のブレーンであった僧、隆光が護持院を建ててもらったが、この寺など直ちに焼失している現在の神田錦町三丁目から一ツ橋二丁目にかけて、幕末まで護持院ケ原と呼ばれた空地などもその一例である。

幕末になるとこの空地帯は、幕府の洋式陸軍の練兵場になったり、はじめ三崎町、のち九段下から移った蕃所調所－洋書調所－開成所－大学南校と何回か名称の変わった、現在の東京大学の法・文・理学部系の前身の学校が建てられるようになった。

明治になるとこの広大な空地は、「洋学祖述」つまり西欧の学問の成果だけを直輸入することを目的とする、官立学校の集中地帯になった。この地区に設立された官立学校を多少のズレを無視して列挙すると、大学南校（開成学校、大学予備間）－東京大学、東京外語大学－東京外語大学、高等商業学校－一橋大学、学習院（当時は官立）－学習院大学、体操伝習所、陸軍大学などが設立された地帯であった。

以上はこの地域に眠る過去の歴史の一駒だが、このような事実からこれらをモチーフに、かつてこの地あるいは界隈にあったであろう雰囲気を抽象して、構成・展開を行った。

完成して、ビルのまわりの東・南・西・北のそれぞれの面の構成と展開が異なった雰囲気のように見えるが、一巡して異和感はなく、東・南・西・北それぞれの眺めが一つにまとまっているような印象を受け、私は成功したことを喜んだ。京王プラザホテル 4 号街路外空間の完成から 7 年を経ている 2 つ目の都市の外部空間である。

私は完成したこのビルまわりを廻っているうちに、1972年初夏のクレタ島の旅のハギア・トリアーダの宮殿遺蹟での空間体験が、このような空間を創り出させたのであろうか、と考えた。

1978 年
用石の種類
門柱（石積）　　　　新小松石切石
石　積　　　　　　　芦野石切石、芦野石雑石
結界表示石　　　　　芦野石板石
犬走り敷石　　　　　稲田石板石
〃　縁石　　　　　　稲田石切石
客溜外壁　　　　　　〃
舗　床 1　　　　　　稲田石板石
〃　2　　　　　　　塩山石ピンコロ
飾　石　　　　　　　スウェーデンみかげ石エメラルドグリーン
館名石　　　　　　　小松石
水の様態　　　　　　流れ

※詳細図はディテール図面編の 171〜172P 参照

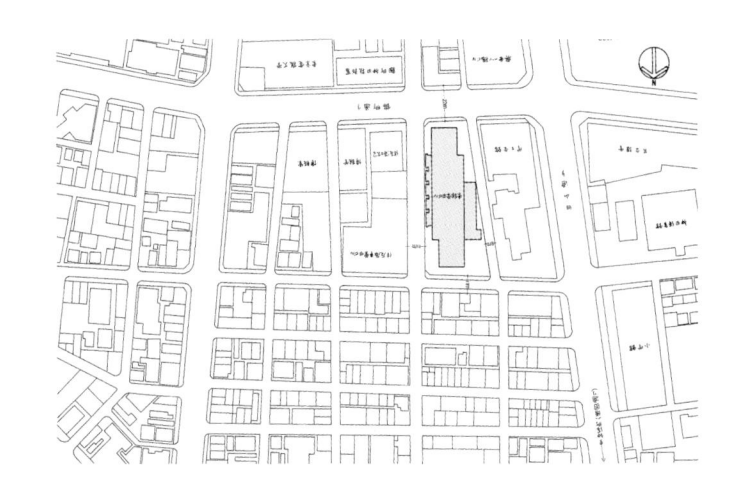

Togin Kurita Bldg. (the present is Hitotsubashi-SI-Bldg.) *'Gai-Kukan'* 1978

The Togin Kurita Building was built on a lot surrounded on all sides by one-way streets, as illustrated, and completed in December 1978. It is a 10 storey building with a cast aluminum façade and a little space was left around the building on all sides.

The *'Gai-Kukan'* of the Togin Kurita Building cuts into a public area in the Kanda-Hitotsubashi business redevelopment area, and succeeds in unifying the road with the façade and premises.

Each space offers a different view to visitors there. The view out from the north entrance especially works well with the openness, as the entrance itself is available to the public in the form of pilotis, which is unique to the area where public spaces are in shortage.

Detailed study has been carried on for each space to give visual tension, so it's a good showcase in that regard. This area was prone to conflagration in the Edo era; therefore quite a sizable fire prevention area was created in the lowland from the current, Yasukuni-dori to the Horibata by the Imperial Palace, the Kanda-bashi Bridge to the Ogawa-cho intersection to the east, and Kudanshita to the west. Ryuko, a monk who was the brain behind Tsunayoshi, the 5th Shogun of Tokugawa of the Genroku period (1688-1703) had the Shogun build the Goji-in Temple. However it burned down after a short while, and was called the Field of Goji-in until the end of the Edo era becoming one example of such an empty plot. The area is recognized as the present Kanda-Nishiki-cho 3 chome and Hitotsubashi 2 chome.

The use of the empty field changed as time progressed towards the end of the Edo era; It was a shogunate-owned western-style military training center for a time, and one part, changed names several times. For example, at first the area named Bansho-shirabesho moved from Misaki-cho(lately Kudanshita) to Yousho-shirabesho next Kaisei-sho , lately Daigaku-nanko (all of a public school of western –studies, lately Literature and Science of the University of Tokyo).

In the Meiji era, this vast plot played a role of 'Yogakusojutsu', in other words, an intensive building site for public institutes and schools that aimed to import exclusively western academics and achievements. Ignoring minor details, the list of the schools built in the area would be: Daigaku Nanko (Kaisei school, a prep school), University of Tokyo, Tokyo University of Foreign Studies and its high school, Hitotsubashi University, Gakushuin University (it was a public school at the time), Gakushuin high school, the Sports Science Institute, and the military academy.

The above are fragments of the sleeping local history, but I picked a motif from these facts to abstract the atmosphere of the place to enhance the design and development.

On completion, I walk around the entire site and look at all the spaces to the east, south, north, and west of the building. While they looks like different atmosphere, I do not feel any lack of harmony but rather see them all as a whole unified landscape. I am glad that this space is a success in that it was the second urban *'Gai-Kukan'* (connecting road and building) that I created, seven years after Keio Plaza Hotel.

There is a rich history to this area that I thought about while I was designing these spaces but while I was walking around the completed project one day I couldn't help but wonder if my impetus for this was really the palace at Hagia Triada on the island of crete that I visited in the summer of 1972.

1978
List of stones
Gatepost (Stone walls): Shinkomatsu-ishi, cut.
Stone walls: Ashino-ishi, cut. Ashino-ishi, fragments.
Stone for boundary lines: Ashino-ishi, flag stone.
Inubashiri stone paving: Inada-ishi, flag stone.
Inubashiri stone curb: Inada-ishi, cut.
Outside wall of waiting room: same as above.
Paving1: Inada-ishi, flag stone.
Paving2: Enzan-ishi, granite setts.
Ornamental stone: Swedish granite, emerald green.
House Name plate stone: Komatsu-ishi.
Water features: Stream.

※ details 171-172p

NTT 広島仁保ビル 外空間 1980
NTT Hiroshima Niho Bldg. *Gai-Kukan* 1980

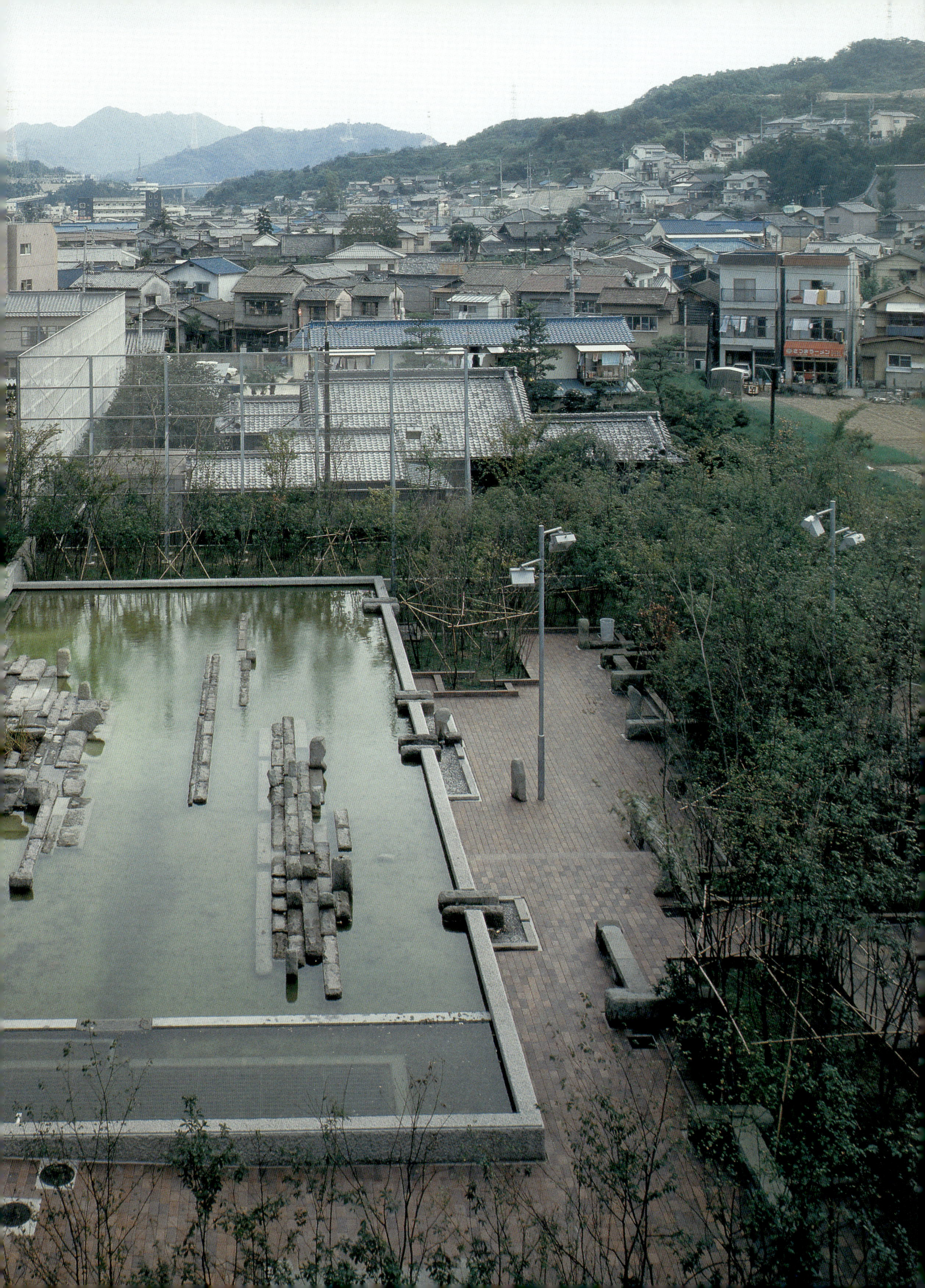

88

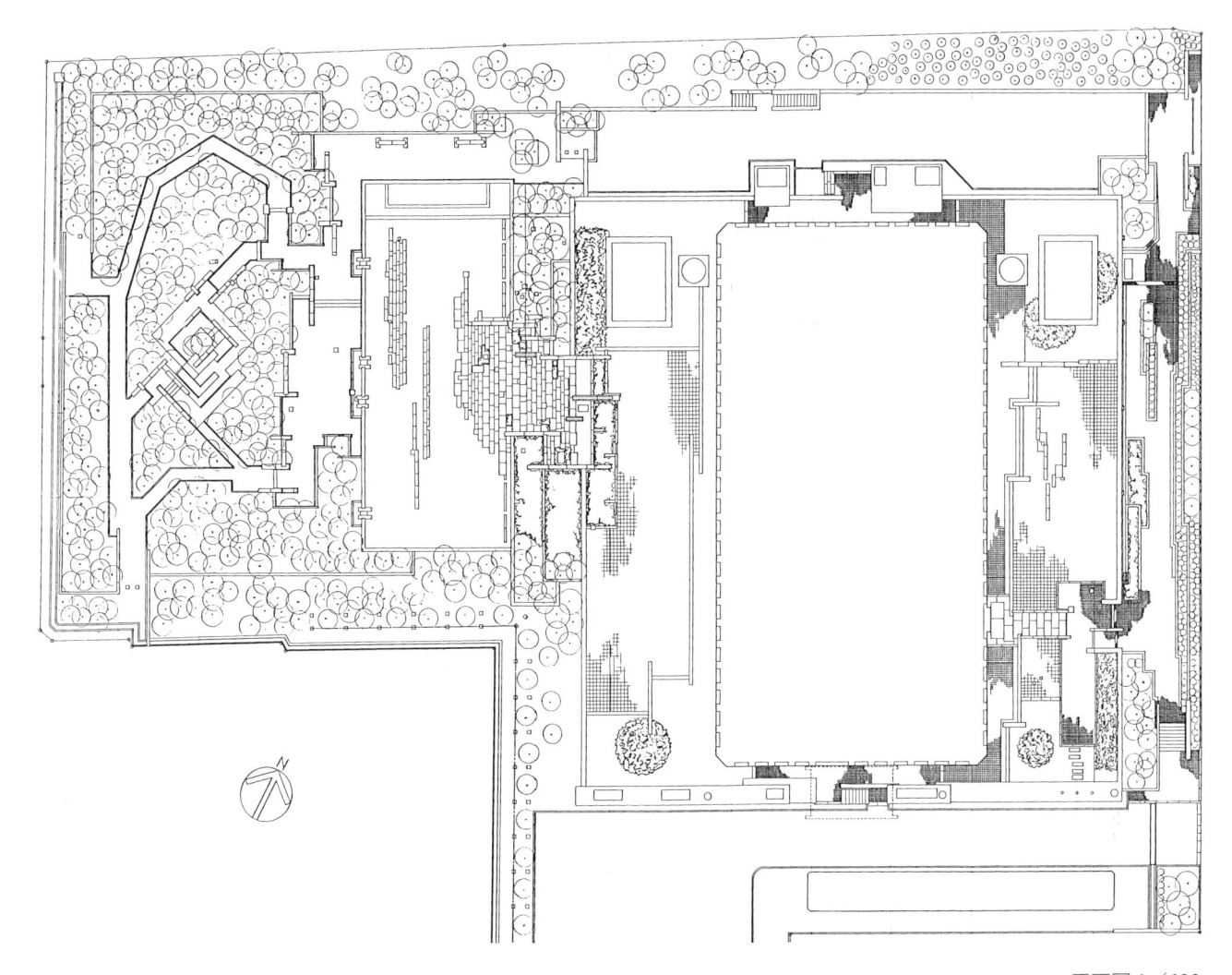

平面図 1/600

89

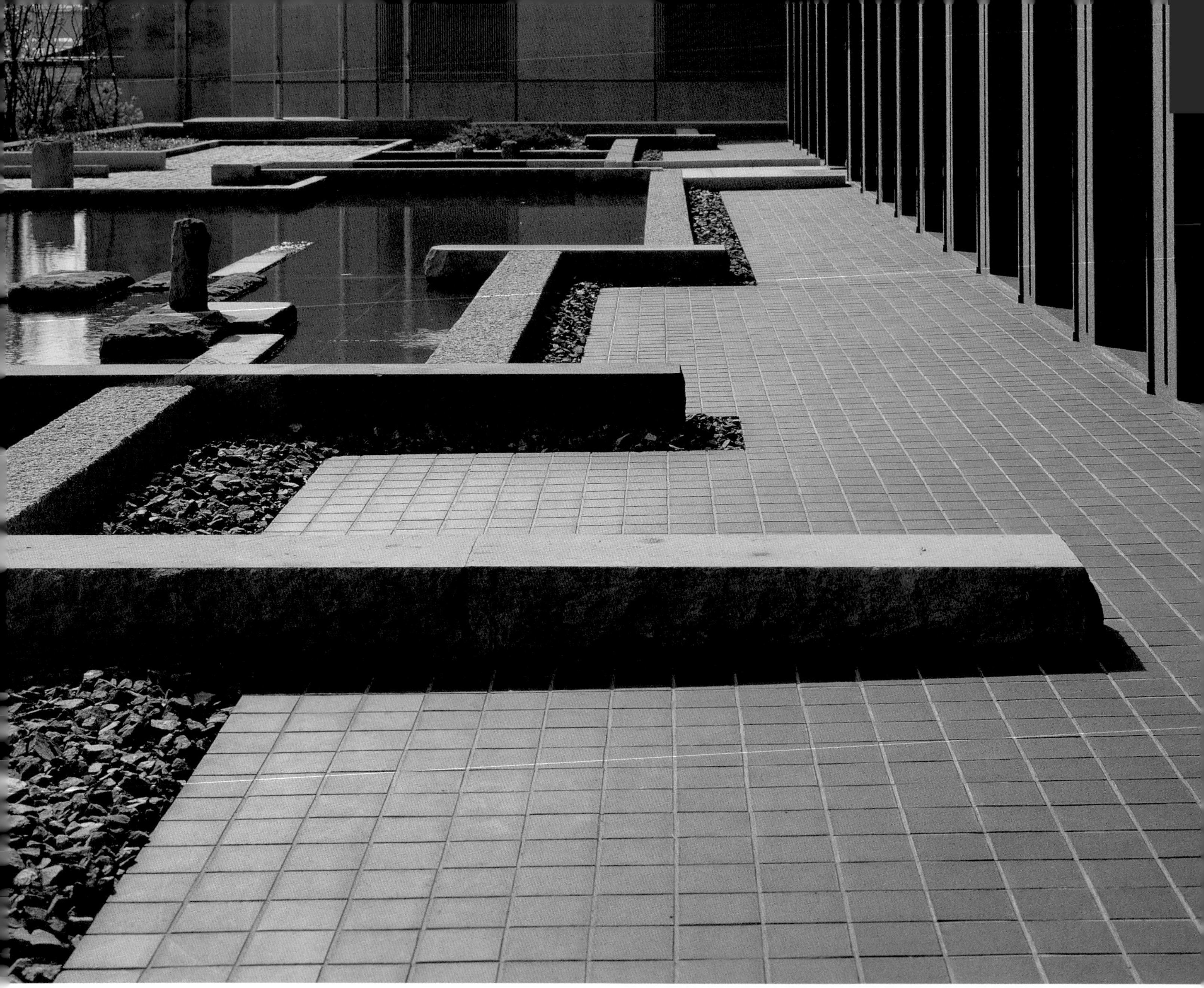

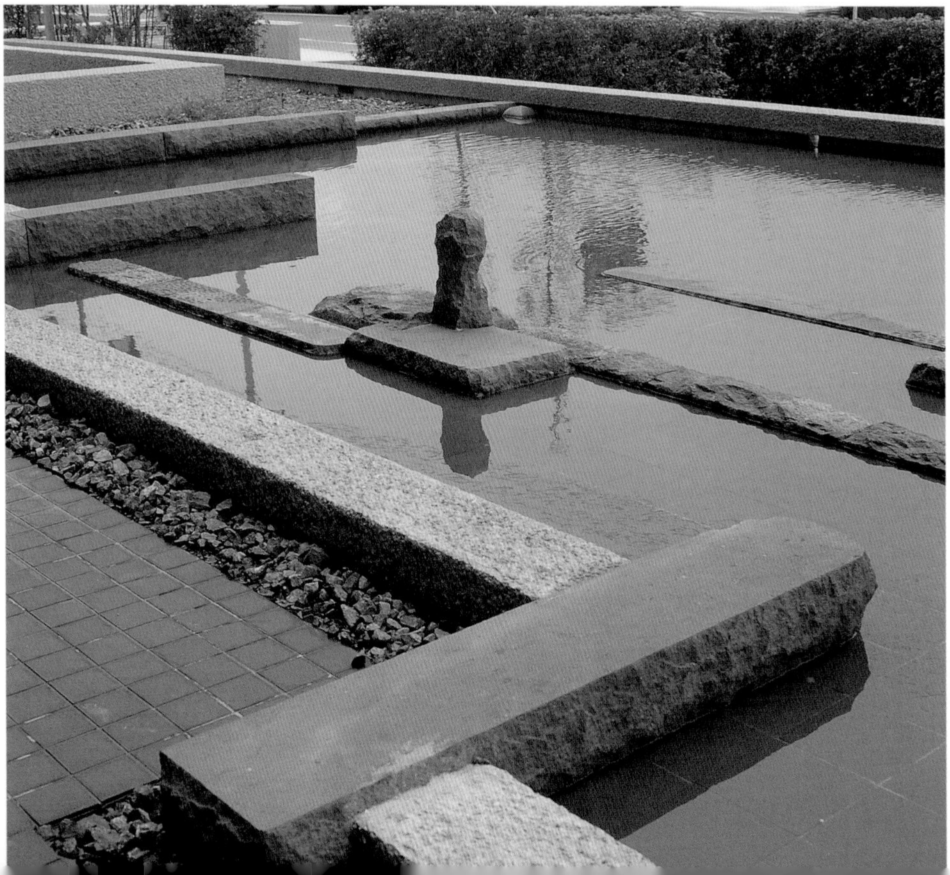

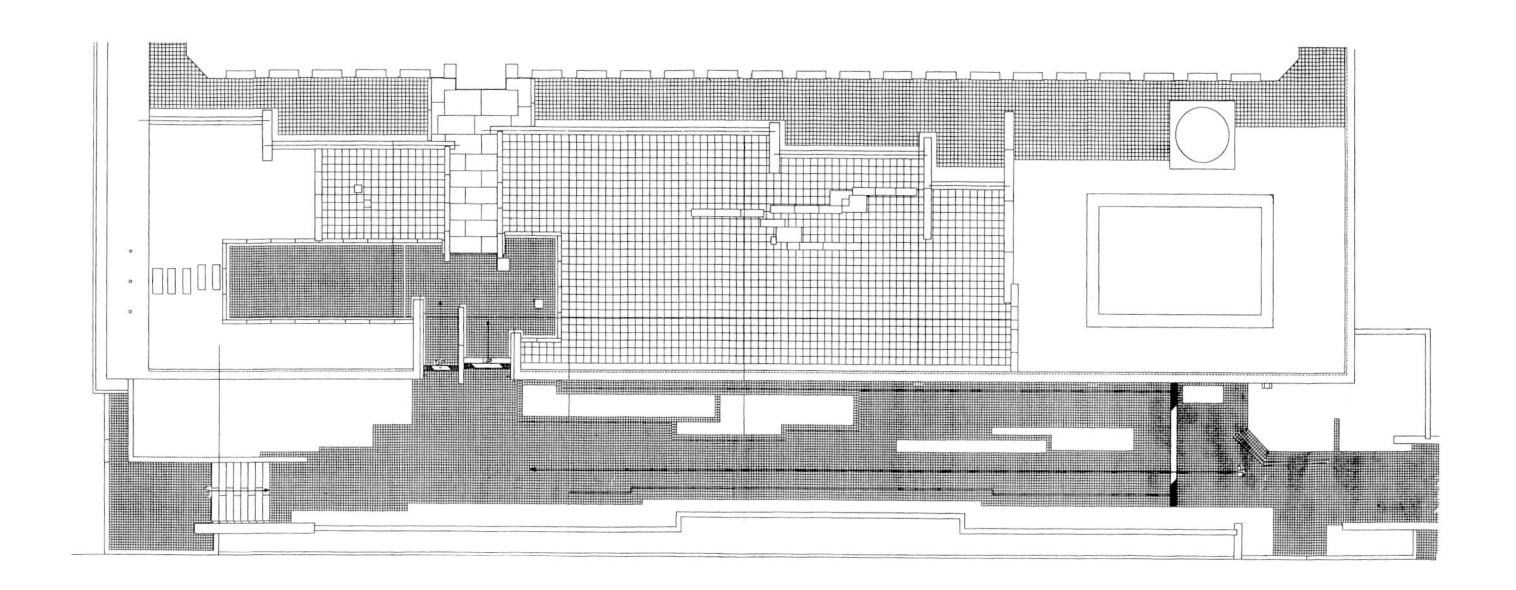

平面図 1/400

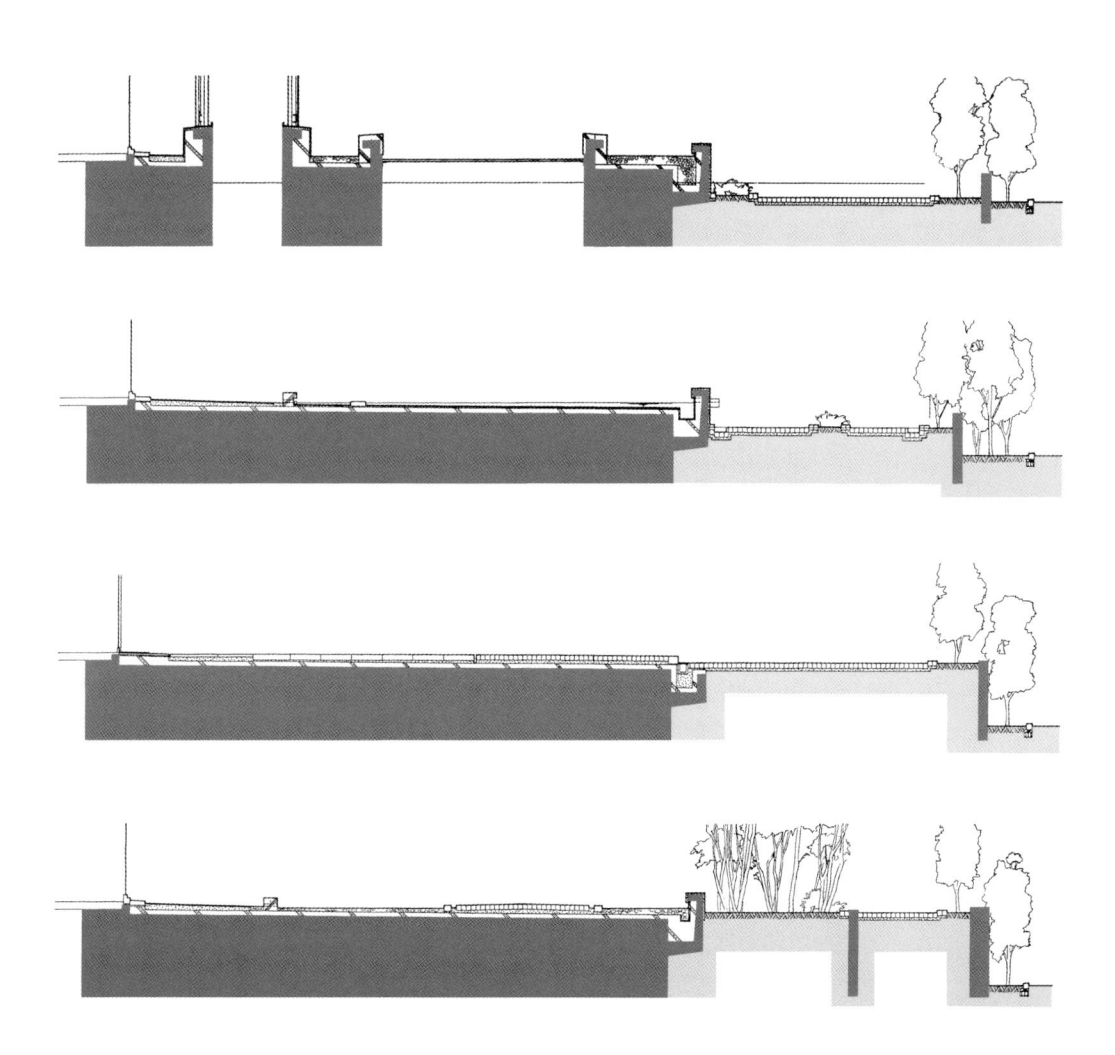

断面図 1/150

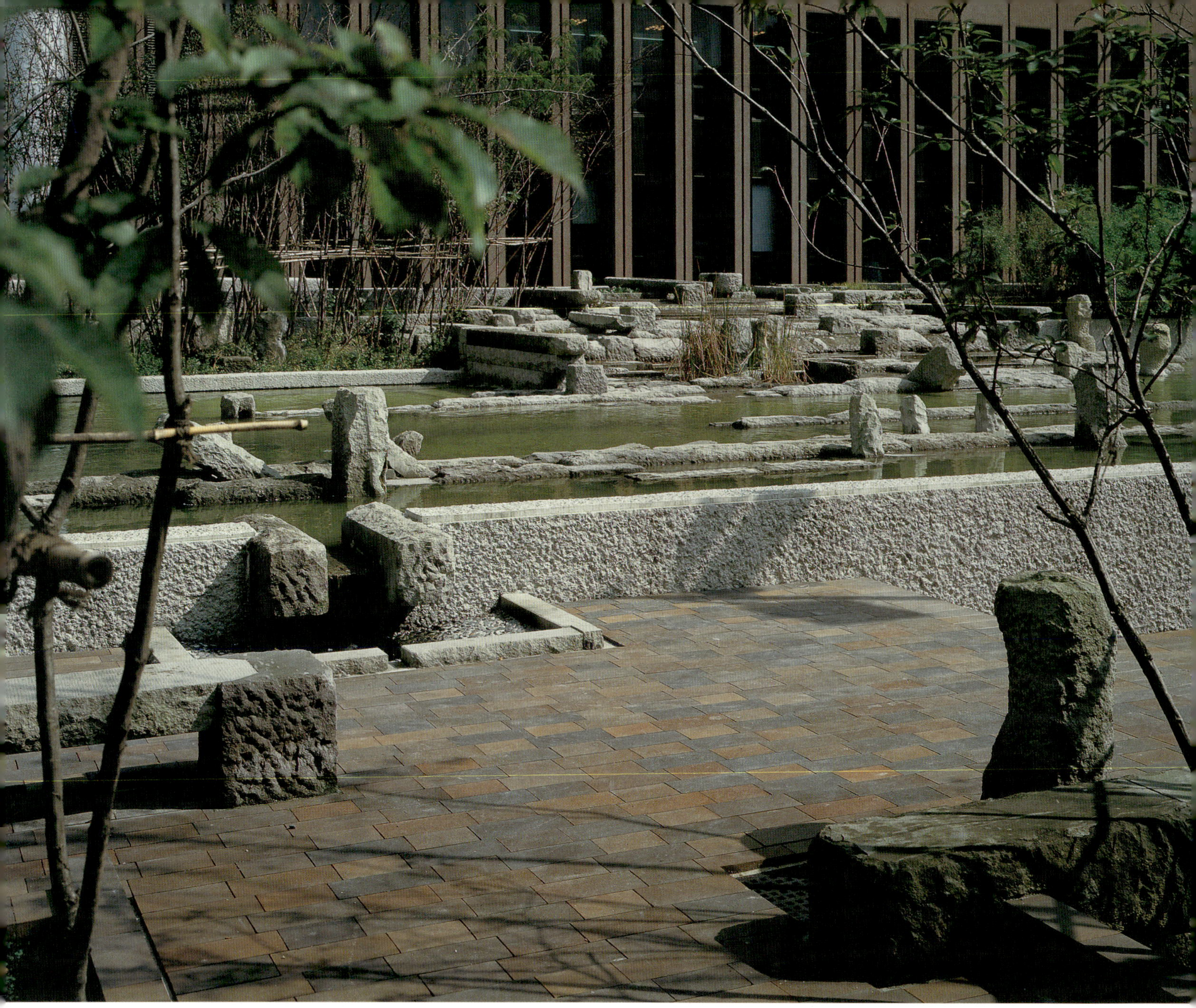

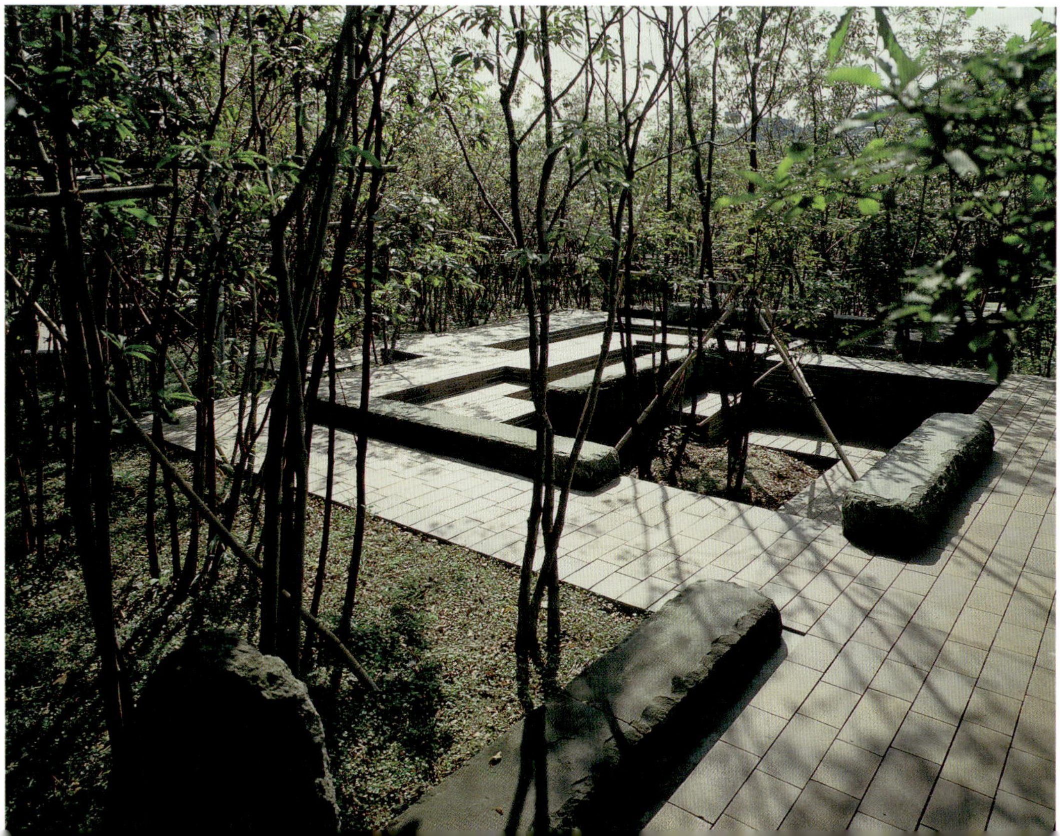

平面図 1/300

断面図 1/150

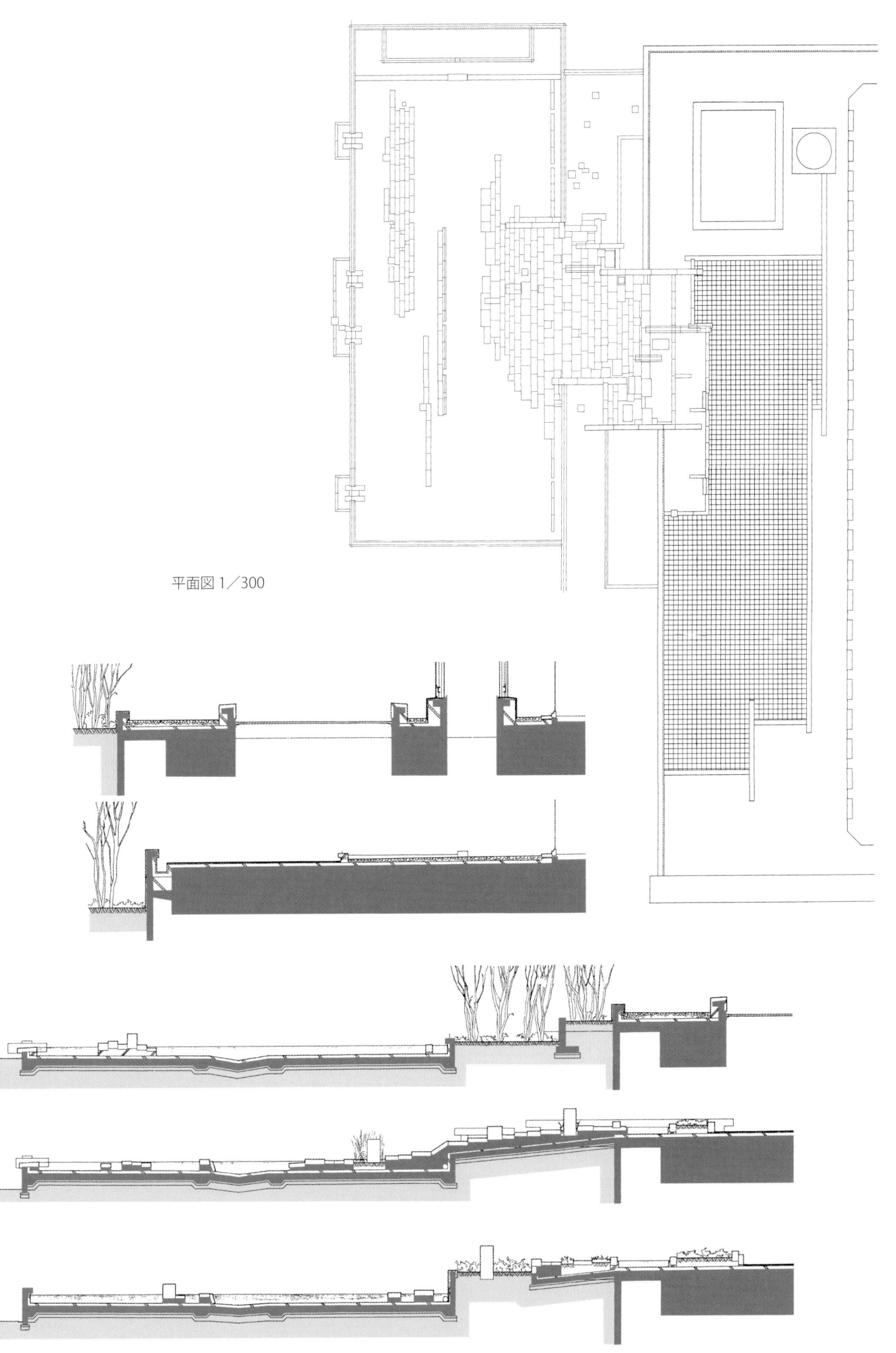

NTT広島仁保ビル 外空間 1980

　国道2号線を下り、広島市内の東を流れる猿猴川に架かる黄金橋を渡って間もなく、南区仁保町二丁目の「NTT広島仁保ビル」の前に至る。辺りは、ビル東側（国道2号線上り側）にマツダ渕崎工場、ビル北側は、水位調整運河を手前に対岸にマツダ教育センターを除くほかは、概ね住宅地である。

　国道2号線に沿って西側に建つこのビルは、地上2階まで有窓、それより上の階から最上階まで無窓に近い地上80mの超高層ビルである。外装を茶色いコンクリートのカーテンウォールによった地上80mの建物は、大衆の視覚からの表情は茶色いBig Boxというところであろう。

　この外空間は、超高層局舎建設によって生まれた総合設計制度による公開空地を、一般に常識のようになっている公開空地処理を超えて、都市環境を構成する外部空間としての主張を構成・展開して、地方文化の発展に一石を投じようという、当時の日本電電公社の広島仁保電電ビル設計プロジェクトの考え方であった。

　要約すると、総合設計制度によって生まれる従来の公開空地の内容を越えて、超高層ビルを効果的な背景として人びとを快適に抱き込むような空間を生み出すことによって、都市環境としての一歩進んだ公開空地ということをアピールしよう、ということであろう。

　既に京王プラザホテル4号街路外空間で話したように、超高層ビルの外部空間を計画する折に最初に検討すべき問題は、"超高層ビルの存在"によってビルの外部に介在する人びとが受ける超高層ビル故の威圧感、微気象故の違和感からの解放をどうするかであろう。

　そこで、建物を東側と西側とに分け、人工地盤を中心に東側アプローチ、人工地盤を直線構成で、西側は長方形によって3～4分割して人工地盤部、水部、舗床部、林部とし、それぞれを有機的に結ぶことによって一つの物語を展開した。物語とは、自然の輪廻がテーマで、有機とは、視覚的にナイーブに人の心に影響を与えようとする為であり、その触媒として水、石、コンクリートフィニッシュ（打放しツツキ仕上）、焼陶板、雑木林等で構成された空間連続を意味する。

　直線による構成は非相称構成とし、動的均衡を保ちながらその直線によって分割され、それぞれの高低差によって幾つかの層に分けられた敷地は、人工地盤を中心に東・西に長く、南・北の方向に短く前下りの遠近感を保つように配慮した。これは逆方向からも視覚的にかなり長い距離感を生んでいる。ことに建物西面から敷地境界までの直線距離約57mは、視覚的に200m程の距離感を生んでいる。

●細部

　人工地盤部：東・西部共、石・コンクリートを素に構成・展開を行い、展開によって生ずる間をそれぞれ水・砂利などの素材によって演出させ、コンポジションと、水と砂利との間に生まれる静と動によってアイレベルを低く絞り、建物の外部空間に介在する人びとが超高層ビル故に受ける威圧感、微気象（ビル風など）故に感じる違和感からの解放をテーマとした。

　ラインコンポジションは建築設計上変更不能なドライエリアの立ち上がりを大きく意識して大胆に展開したが、ここでは成功している。これは人工地盤の広さと、建物をBig Boxとして意識したことにもよるのであろう。演出素材として用いられた水は、ここではその動的な効果もさることながら、巨大ビルによる都市温度の上昇、降雨の処理のデモンストレーションであり、且つ積極的に使われない防水層の保護への一つの方向の挑戦でもある。

●水部

　自然の様態の中で水は主役である。このことは"水は自然を涵養してきた"というよく使われる表現からも明らかなように、自然界の輪廻の根元である。林部に展開した雑木の樹林涵養のため、人工地盤上の雨水及び冷却塔よりの排水の貯水と、これらの水を林内に調整放水する遊水池として成り立つ池は、他方、人工地盤部と水部とを有機的に結ぶ物語の大きなポイントでもある。

●林部

　物語のエピローグの「自然な自然さ」溢れる処である。南の方向に遠景として目に入る黄金山の樹林を構成している雑木の林に、心の風景としてつなげようと、雑木林とした。

　林部には散策の為の道を通し、林内に小児が自然を相

手に遊ぶであろう点景ともなる遊び場を設けた。

　今は亡き東京芸術大学教授であられた山本学治先生が、生前、誠文堂新光社発行の『外空間－原風景への思惟』(深谷光軌著・村井修撮影、1975年)の書評を『商店建築』7月号(1975年)に「『外空間－原風景への思惟』現代外部空間に生きる"自然を愛する心"と"逞しい異質性"」と題して、亡くなる少し前であったのであろうか、載せられた。私は、山本先生には生前一度もお目に掛かったことはない。ということは私がもの創りであって、はにかみやであったからか。ところが1976年、人の紹介で北海道銀行創始者の島村融氏の墓を宇治平等院の墓地に創る依頼を受けた。私は寺の生まれで、東叡山上野寛永寺子院等覚院の出身の在家僧侶である。そして等覚院の先々代住職の薗光轍はかつて宇治平等院の抄門で、後の京都妙法院の門跡であったこともあり、私の一族である。

　私はこの奇遇ともいうべきめぐり合わせに驚いた。島村融氏は無神論者であったから。私なりの無神論者の墓碑その他を建立した。2年後、島村さんの紹介で親戚の山村さんから、生前墓碑その他の注文を受けた。平等院の島村さんのお墓が気に入ったから私の処もとのお話で、無神論者・山村さんのお墓を東京都営の谷中霊園に建立することになり、墓碑をどのような形にするか迷った末に、谷中霊園、寛永寺霊園の墓碑を見て廻った。その折、寛永寺霊園の中で先生の墓碑にめぐり合い、この奇遇に驚愕した。というのは、この1年前に先生は亡くなっておられたからである。他方、なんと山本家は寛永寺一山の何処かの子院の檀家であったのだ。

　その書評の中で、「期待される都市空間への進出」のタイトルで次のように述べられている。

　「これまでの深谷さんの作品には、京王プラザを除けば都市の公共空間に割り込んだものはない。ほとんど敷地のなかにとり込まれた外空間である。けれどもはじめに書いた理由で、私は深谷さんの外空間がむしろ個々の敷地や建物を背景とし道と一体化することを、そして現代の都市空間のなかに新しい形と人のかかわりを持ち込むことを期待したい。それだけの表現力の強さと意味深さがあるからだ」と。

　私は1970年以来、私の創る外部空間を外空間と名付け、それを創ることに生涯をかけようと歩んできたけれど、ともすれば挫折へ心が傾く時の流れが訪れる日々もあった。そういう歩みの日々の中に、先生の書評に励まされながら明日に望みをつないでいた。そういう時の流れの中で、1978年、先生が生前「個々の敷地や建物を背景とし道と一体化して、現代の都市空間のなかに新しい形と人のかかわりを期待したい」と言われた。その期待通りの外空間、1978年東銀栗田ビル外空間が生まれ、3年後1981年NTT広島仁保ビル外空間が生まれた。

　もし先生の外空間への励ましと理解の文章を『商店建築』に見出すことがなかったら、これらの2つの空間は生まれなかったろう。

　1975年サンかがや外空間以来幾つかの石の造形物を配して一つの風景としてつなぎ、将来空間に介在する人びととの間に、より深い感動の美が生まれることを願った。その2つの空間は、東銀栗田ビル外空間、NTT広島仁保ビル外空間は、芦野石を素材とした部分の造形物は、共に制作以前の図面はない。ということは共に現場でのハプニングによっているからである。これは素材の芦野石が現場のハプニングによる制作を容易にしたことが、制作の場の雰囲気に即したハプニングを可能にしたからであろう。

　東銀栗田ビル外空間については既に述べた。NTT広島仁保ビル外空間の現場制作は1980年に始められた。この空間では、芦野石を造形素材としたハプニングによる造形空間は、97頁以降の写真に展開されている。その5つの空間に展開したモチーフは、現場制作着工時に宿泊していた上八丁堀の広島グランドホテル隣地の上幟町に立地する旧浅野公庭園の縮景図の辺り一面に漂う阿鼻叫喚の巷の様が彷彿された。この様子と現地の物売りの老婆の話を取り混ぜて、亡き原爆の犠牲者の鎮魂を含んで、彼岸と此岸の物語を展開した。

　それは正にこの地に眠る歴史の一時機を掘り起こして、

その痕跡を留めようとした

　以上のような空間創りの手法は、私の石の造形の理解者として唯一人の相棒の福島県二本松出身で、(株)タカタ・高田浩雄氏紹介の石工、宮本貞八君が広島仁保ビルの外空間の仕事を最後に帰らぬ人となったからである。

　そして時代は変わった。

　私が一人で我が道を歩み始めたのは、この彼の死がきっかけとなった。

1986年　工学院大学八王子校舎5号館～11号館群
　　　　外空間
1987年　工学院大学八王子校舎3号館　外空間
1989年　日野自動車工業本社 "日野台の杜" 改修制作
1991年　スコープ本社 GLASS HOUSE　外空間
1991年　三番町 KS ビル　外空間

などの主だった作品が生まれた。

　その一つ一つの作品の場を訪れる度に、彼と共に在った日々に、彼が私に伝えた石の扱いについての諸々が、息づいているのを発見する。そして二人三脚であった楽しい日々をなつかしく想い起こす。

1980年
用石の種類
石　積　　　　　　　　芦野石切石
造型貼石　　　　　　　〃、芦野石板石
ベンチ　　　　　　　　芦野石切石
舗床（アプローチ）　　みかげ石尾立石板石
　　　　　　　　　　　みかげ石北木石ピンコロ
舗床（陶板）　別注
　　　　　　　伊奈製陶　t40×400×200
流れ床（伊奈プレート）　t10×400×400
水の様態　　　　　　　流れ、堀
自動潅水設備　　　　　一部スプレーノズル式

※詳細図はディテール図面編の173～175P参照

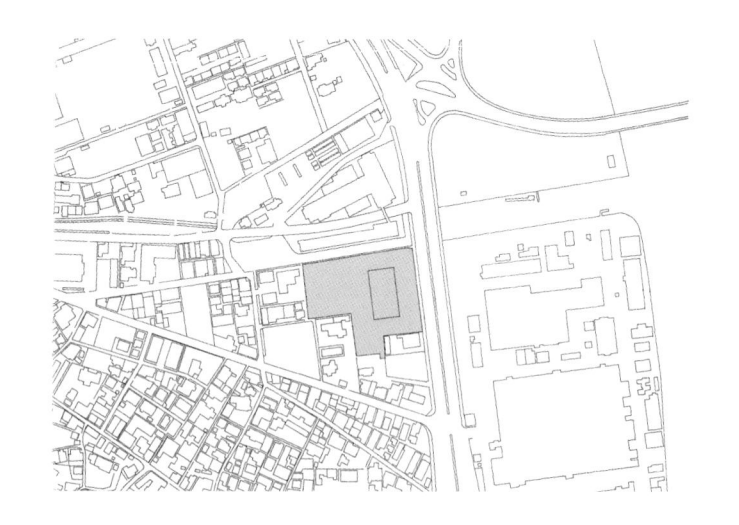

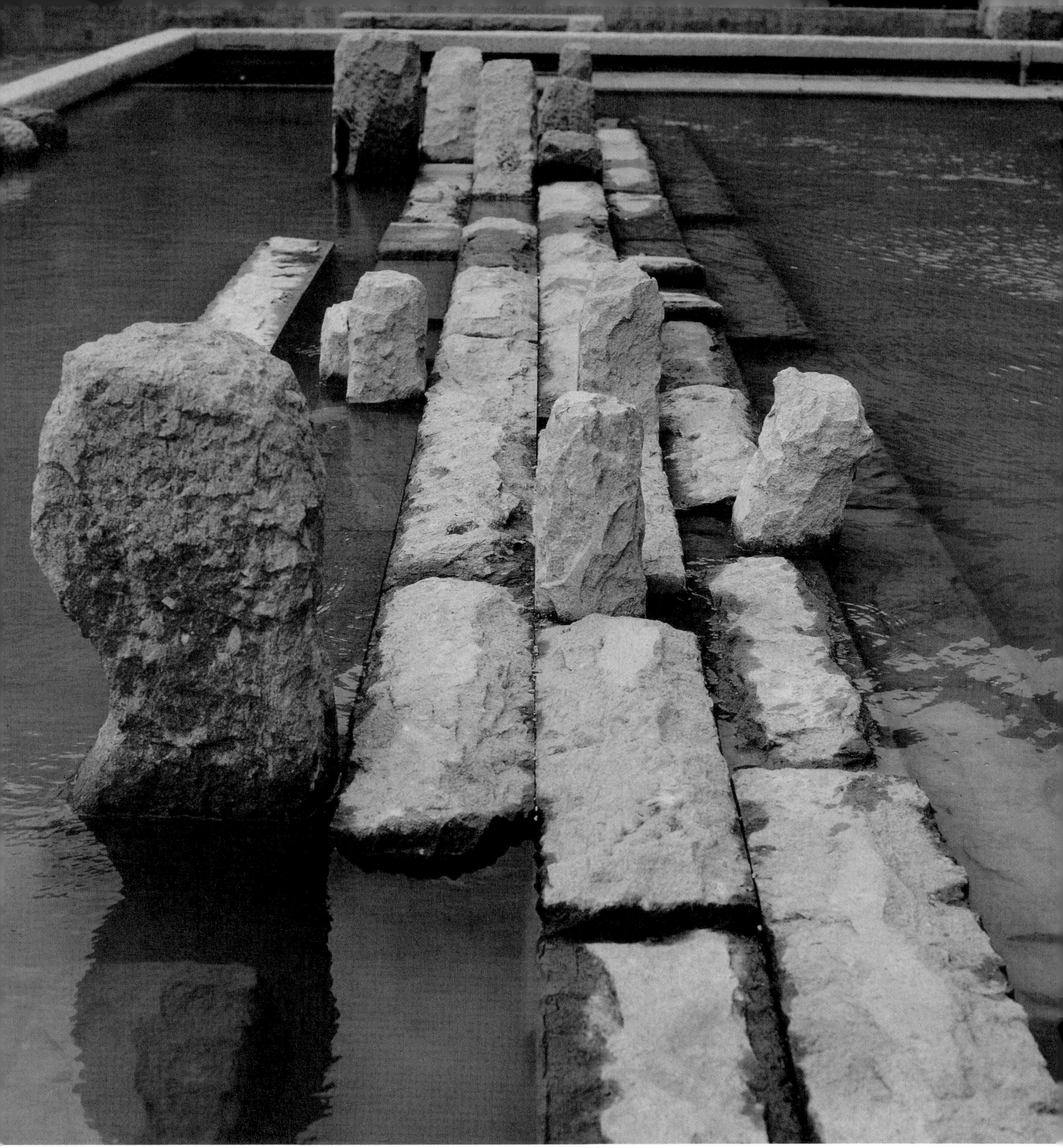

NTT Hiroshima Niho Bldg. 'Gai-Kukan' 1980

Going down 2nd National Road, soon after crossing Ogon-bashi on the Enko River which on the east side of Hiroshima city, arrival around to the front of "NTT Hiroshima Niho Bldg." at 2-chome, Niho-cho, Minami-ku. On the east side of the building (up line of 2nd National Highway) is the Fuchizaki Factory of Mazda Motor Corporation, on the north, the water level adjusting canal is in the foreground, Mazda Motor Corporation Education Center is in the back, mostly residents on the rest.

This building which stands alongside 2nd National Road on the west side is a skyscraper height 80m. Setting windows up 1st and 2nd floor but hardly any windows above to the top floor. The outside walls brown colored concrete curtain walls, it seems the 80m high building looks like a big brown box for the domestic people.

These days according to the regulation regarding by public for the integrated design of very high buildings, a certain amount of open and outside space, should be left in the building site and offered as a public space included in the exterior space.

It was advanced cases that expand the exterior space as the united urban environment composition, aimed glowing the district civilization by project of Hiroshima Niho Denden Bldg. of NTT.

In other words, we wanted to go beyond the conventional way in utilizing the public space produced by the integrated design regulation. We wanted to design a space that forming a space that embraces the people there comfortable, using the surrounding buildings with the skyscraper facades as an effective as a desirable background for create an advanced public space as 'Gai-Kukan' it's contributes to the improvement of the urban environment.

As I mentioned in connection with the Keio Plaza Hotel, Street No.4, 'Gai-Kukan' the problems what should be considered at first is how to relieve the people who live in outside skyscraper building from the discomfort feeling due to the unnatural climate and pressure viewing.

For this I divided the building space, the east side and the west side. The east side placed for approach, placed artificial grounds direct lines in the center, and the west side was divided into 4 components rectangles: the artificial ground, the water section, the paved floor and the woods. It aims to represent a story of natural providence as the four components. As the story, the theme is nature's cycle of life, and by organic composition, Aiming to influence people gentle and delicate through visual sensation by the 'Gai-Kukan' . It would be catalyzed by the unified space of the water, space, concrete finish (cast finished with prodding), porcelain plate, bush, and others.

Composition by direct lines is made asymmetrically, dividing the space into a number of parts, yet keeping a dynamic balance among them. Divided into parts of layers as different heights, its compound spread directions from east to west as long distance and from south to north as short distance. It's made feel more sinking viewing than accrual by perspective in which stand forefront. This arrangement produces a visual sensation of feeling long distance as view in which stand opposite direction. Especially, the actual distance of 57m from the west surface of the building to the border of the compound is appears to the eye to be about 200m.

*Details

Artificial Ground: On both east and west side of it, composition and development was done with stone and concrete, and the gaps generated in the course of development were filled with water and gravel. The eye level was kept low by means of static and dynamical balance between the water and gravel and the rest of the composition. The theme was to relieve the people in the 'Gai-Kukan' of the building from the mental pressure due to the skyscraper and from the feeling from the discomfort feeling due to the unnatural climate like a draft through buildings, and pressure viewing.

Line composition was made daringly, keeping well in mind the vertical surface going up in the dry area, which could not be altered because of the design of the building itself. It was successful, I think. The reason perhaps the artificial ground was made wide enough and the building was treated like as a big box. The water, using as the production of the space, not only to give dynamical effect, but also to restrain temperature rising of the city due to ultra-high buildings, and processing of rain water, and it's indicate the direction of way with protect the waterproof layer.

*Water section

Water is the principal actor in nature. As the quoted expression, "water has fostered nature," indicates, water is the very root of the nature's cycle of life. The reservoir stores the rainwater fell on the artificial ground and the water from the cooling tower, these provide controlled flow of water for the growth of the trees of the bush in the woods. This is an important point that connects, in organic composition, the artificial ground and the water section.

*Woods section

This is the epilogue of a story, so to speak, where it is full of "natural nature." The grove was so arranged that it could be appear to overlap the bush in the forest of Ogonzan, seen in the distance toward south.

The woods in road was built for people to take a walk, and

the playground was made for children to playing with nature, which could also serve as an object in the scenery.

I have already explained the Togin Kurita Bldg. 'Gai-Kukan'
The work on the site of the NTT Hiroshima Niho Bldg. 'Gai-Kukan'. started in 1980. It's case, the 'Gai-Kukan', using happening method, with Ashino-ishi as the material, the photo in p97. The motif is the imagination of the pandemonium of Shukukeien in the garden of old Asano family in Kaminobori-cho, near the Hiroshima Grand Hotel where I stayed at the time at the work, I was imagined the story based the mortal world and the heaven, prayer for the victims of the atomic bomb with the tales by an old sales woman there. I attempt dig up the buried history and keep the trace.

The very techniques above were developed together with my only pal, Sadahachi Miyamoto, who understood the way I produced the stone forms. He is stonemason from Nihonmatsu, Fukushima prefecture, and was introduced by Hiroo Takada Takata Co. Ltd. But he pass away after finishing the job of here.

And the times changed.
His death casted a shadow my mind, gave direction alone way. Every time I visit one of these 'Gai-Kukan', I find the trace with the stones to which he instructed to me in that time. I reminisce that I enjoyed working so closely with him these days.

1980
List of stones
Stone walls: Ashino-ishi, cut.
Surface of a form : same as above. Ashino-ishi, flag stone.
Bench: Ashino-ishi, cut.
Paving (approach): Odate-ishi, granite, flag stone. Kitagi-ishi, granite setts.
Paving(porcelain): Specially ordered.
Ina porcelain: t40×400×200
Bed of water flow (Ina plate): t10×400×400
Water features: Stream, moat.
Automatic irrigation system: Part of spray nozzle.

※ details 173-175p

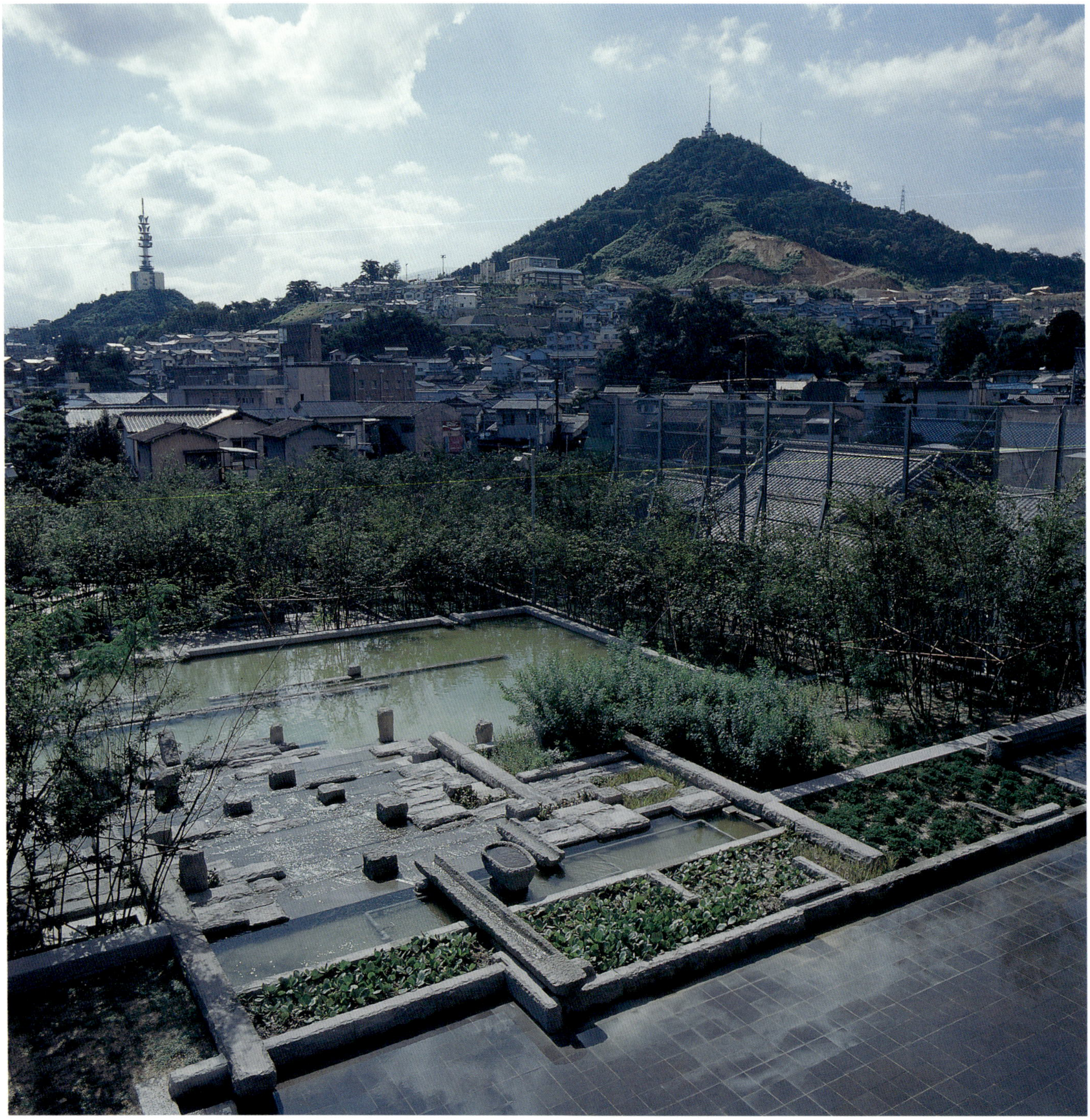

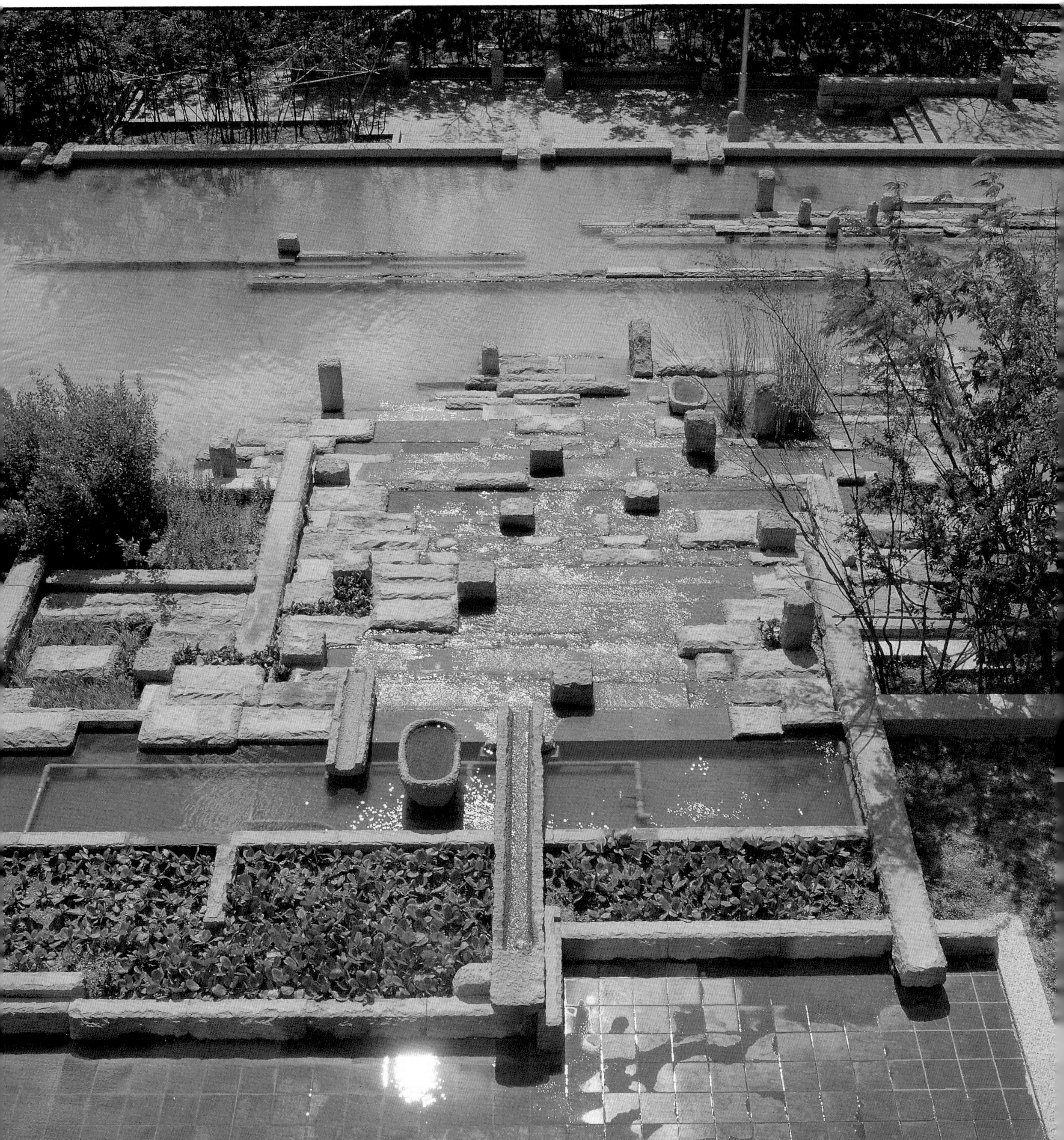

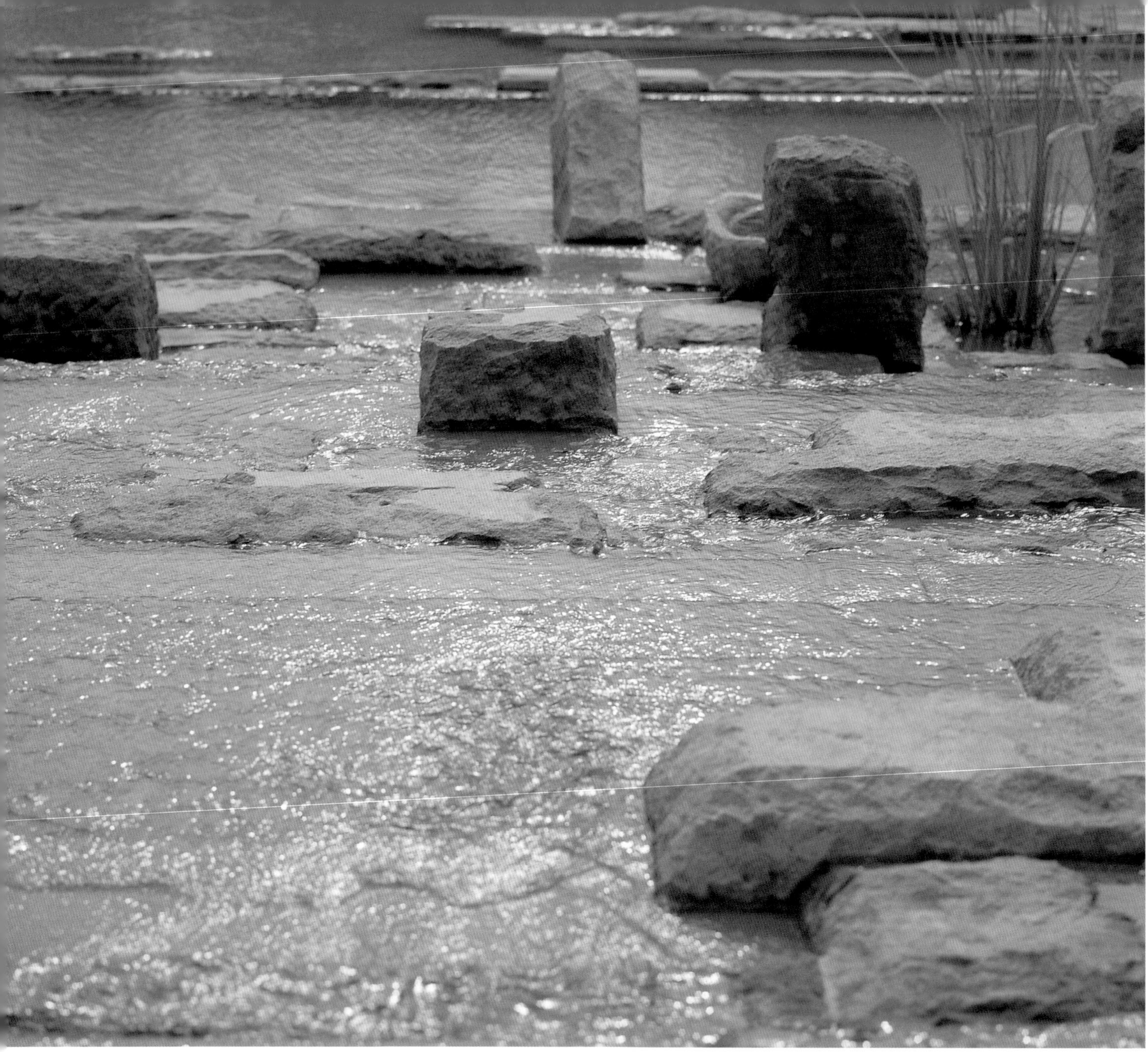

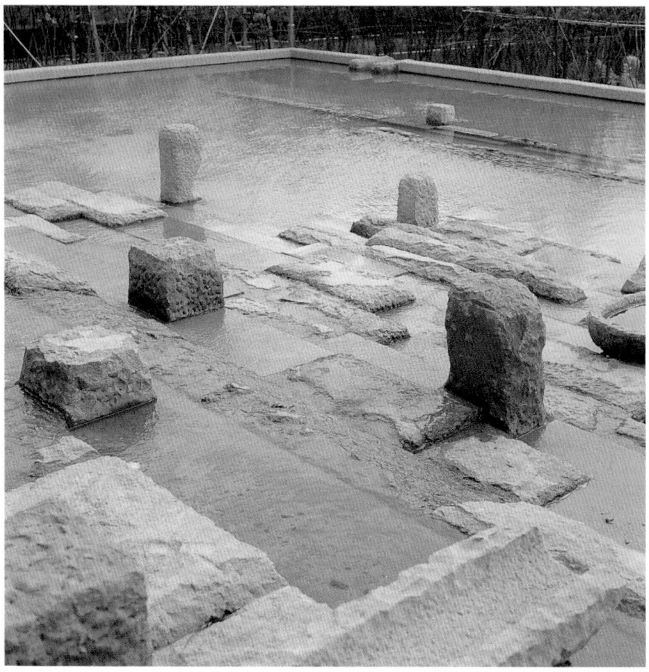

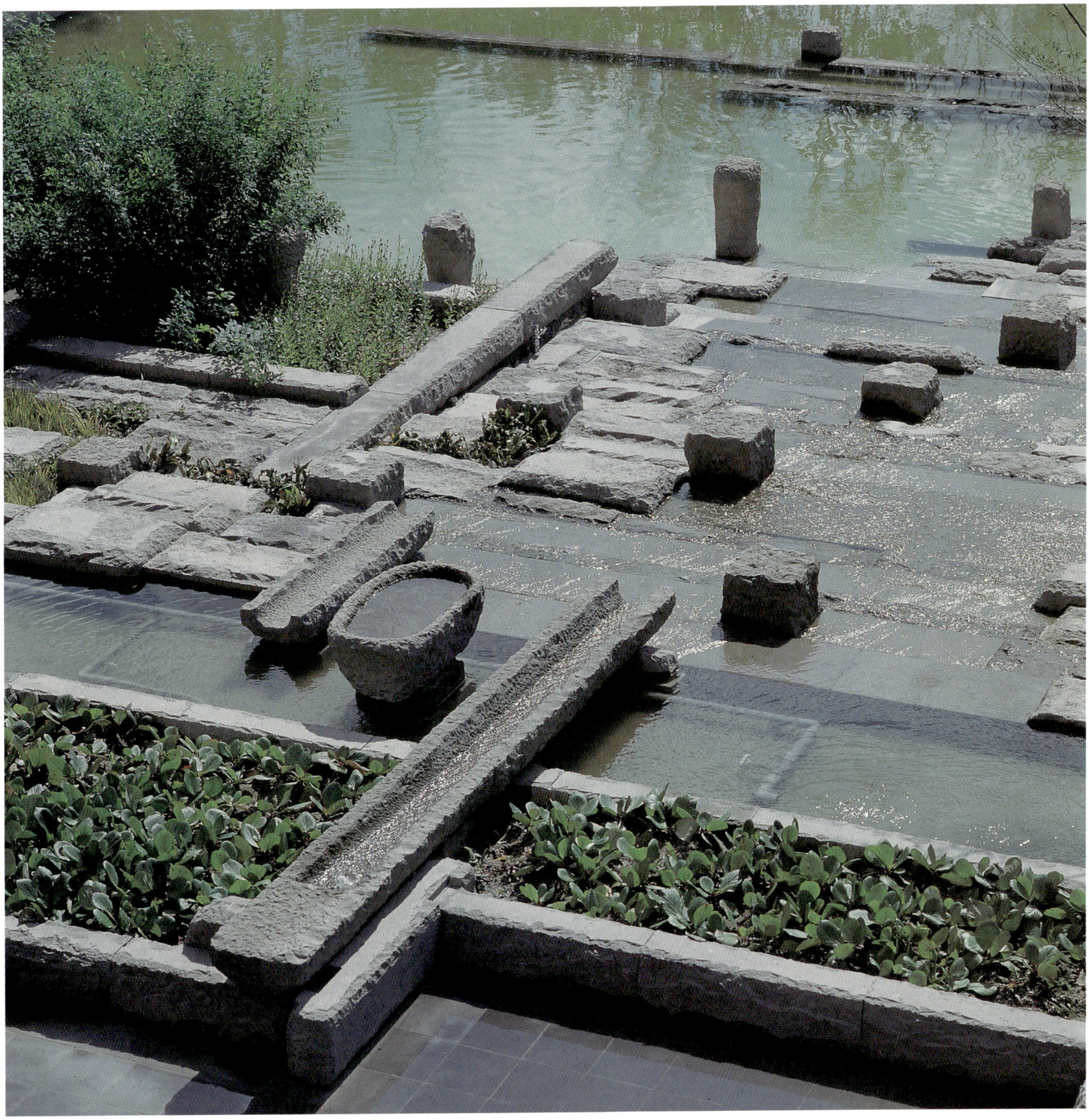

104

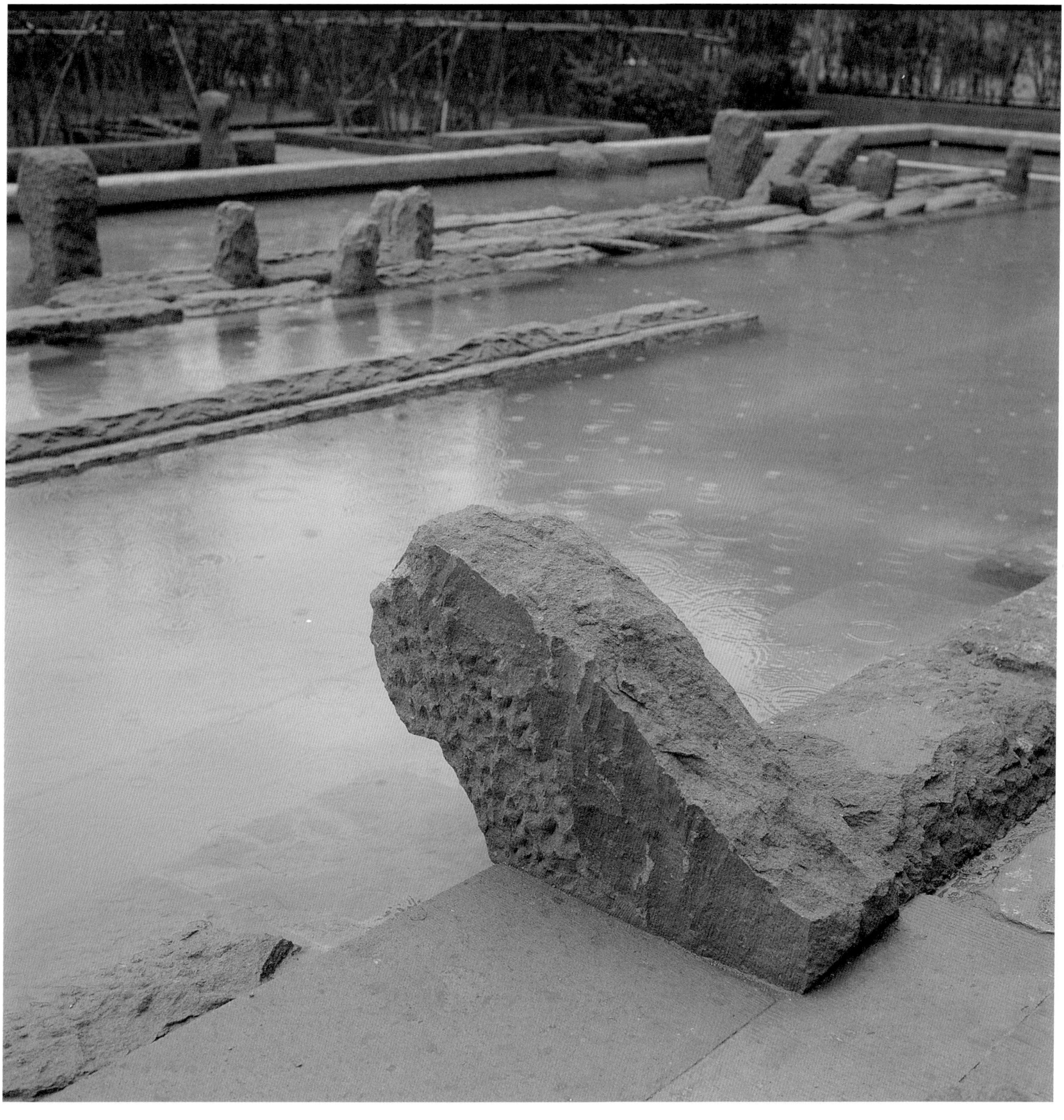

工学院大学八王子校舎 5 号館〜11 号館 外空間 1986
Kogakuin University Hachioji Campus, Bldgs. No.5 to 11 *'Gai-Kukan'* 1986

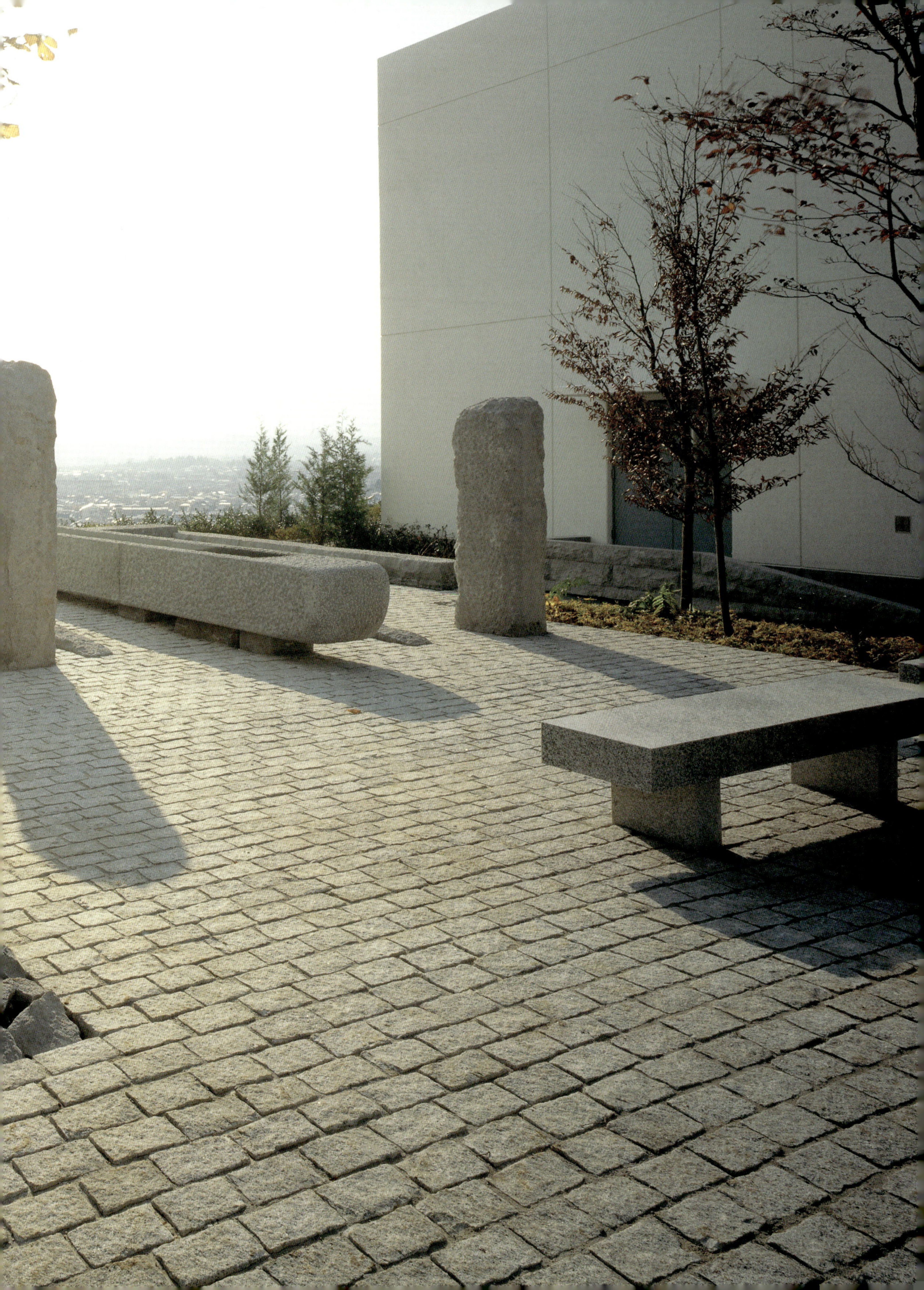

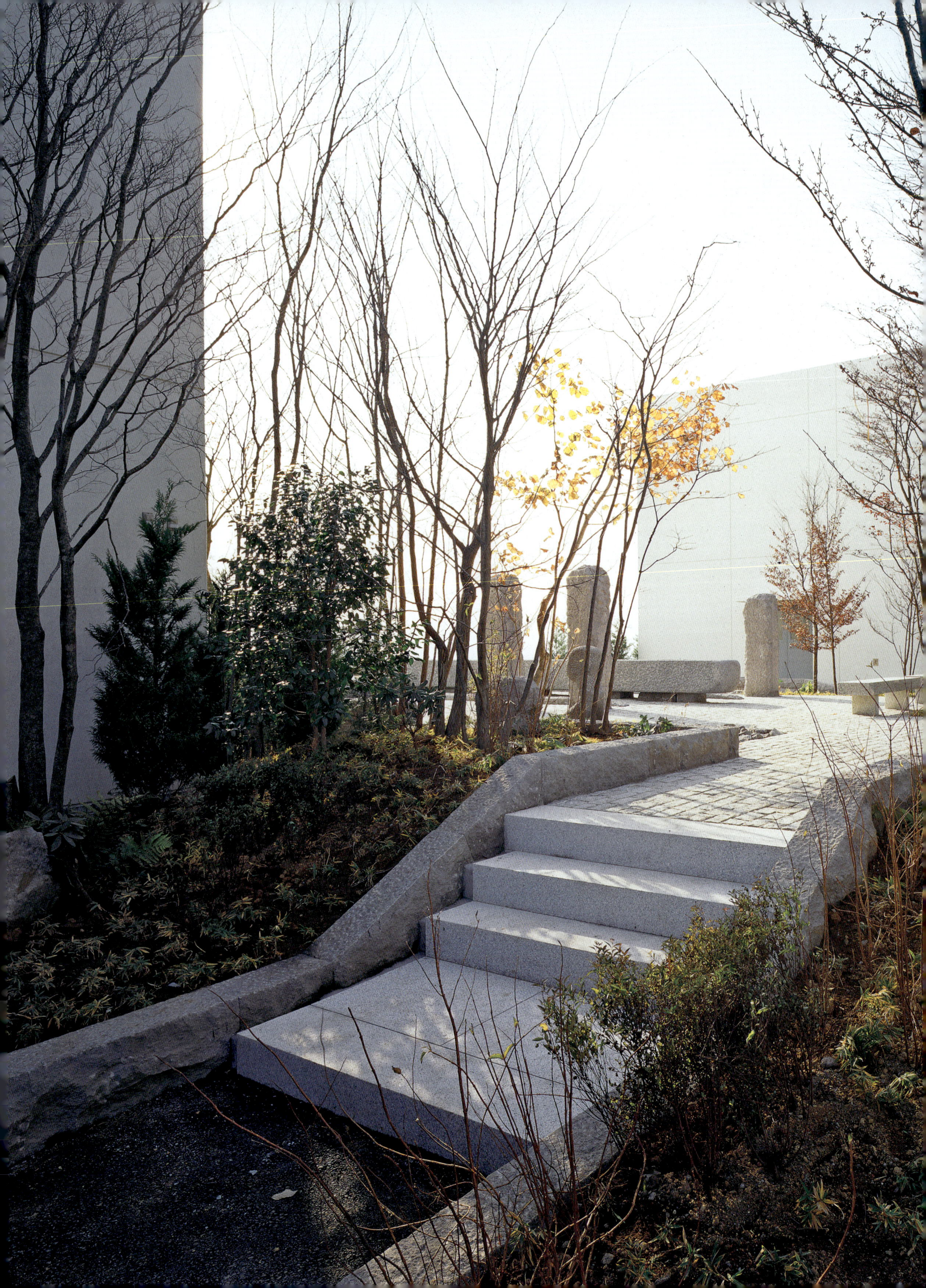

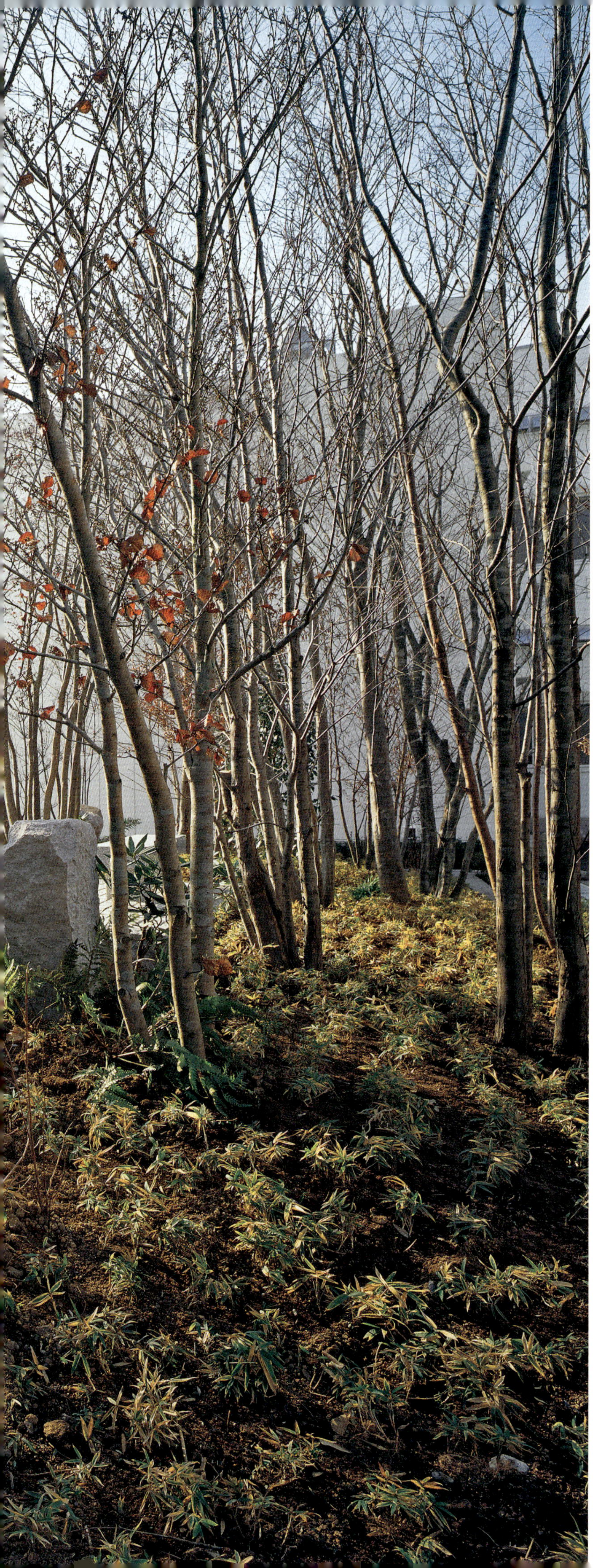

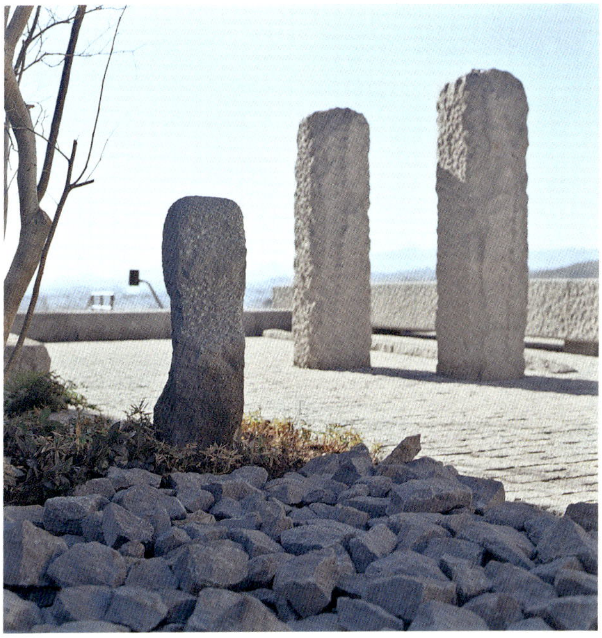

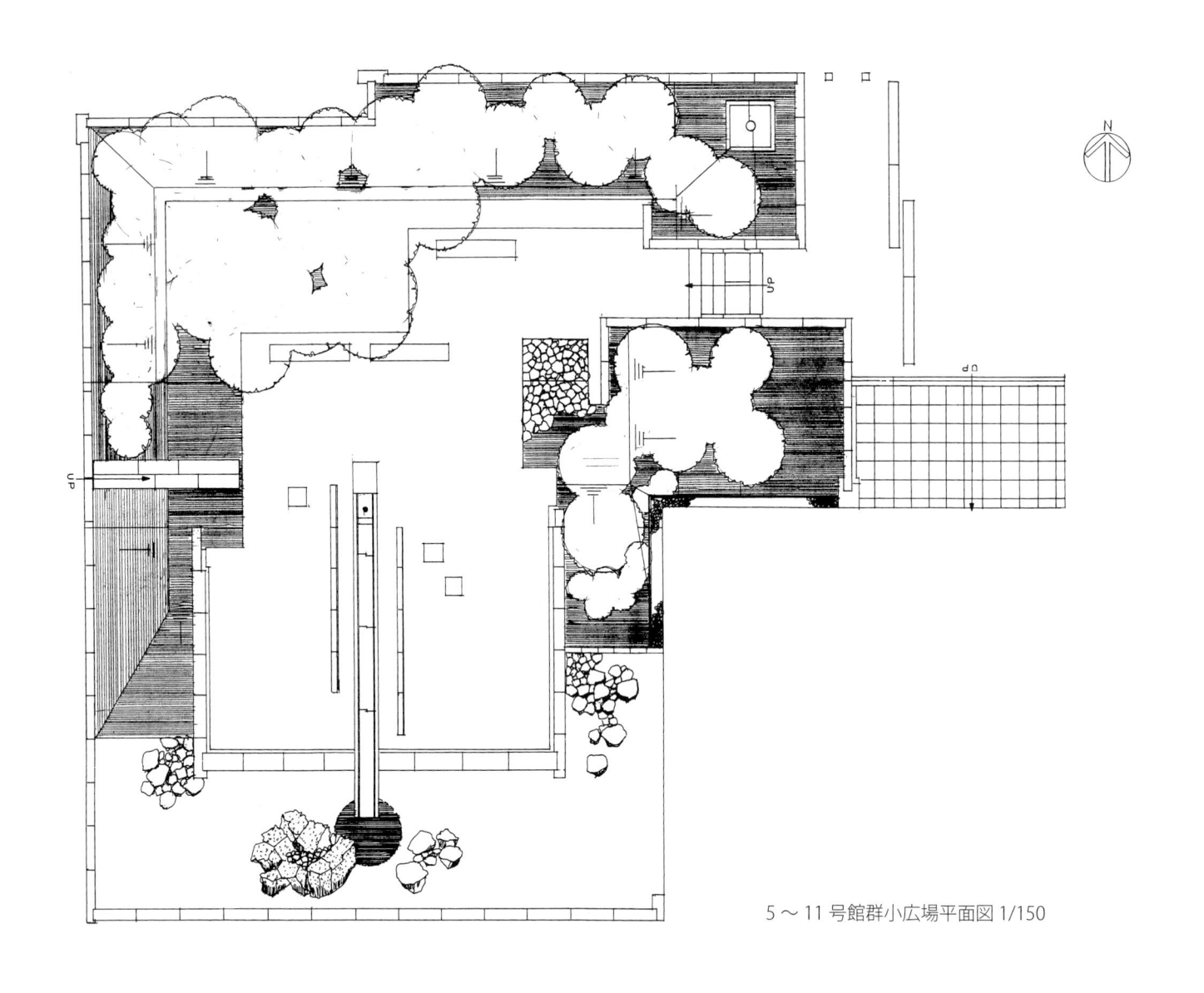

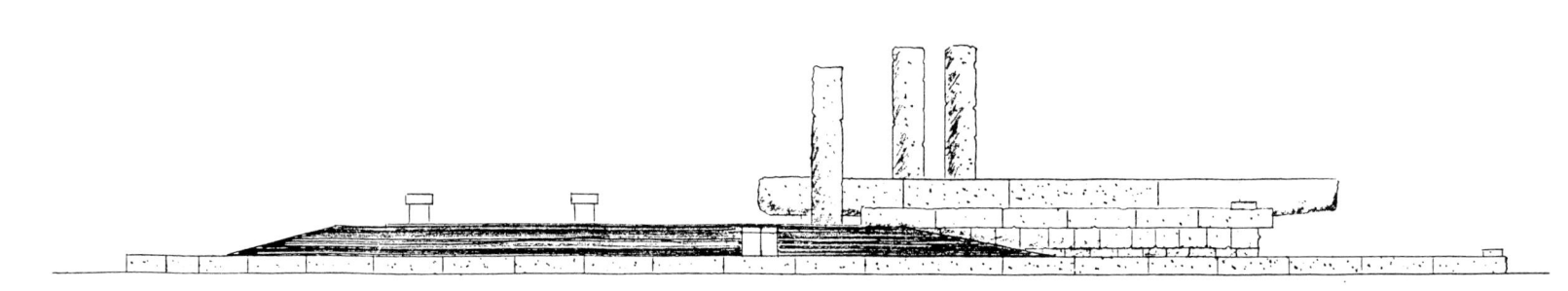

5〜11号館群小広場平面図 1/150

111

工学院大学八王子校舎5号館〜11号館 外空間 1986

　工学院大学八王子校舎5号館〜11号館群外空間、及び3号館外空間の2つの空間は、前者の外空間は5号館〜11号館群の建築群の完成に相次いで、1986年に完成。後者の外空間は、既存の3号館の建物の完成後、1987年に完成した。

　工学院大学八王子校舎の立地する辺りは公道に面した正門から5号館〜11号館群敷地の台地に至るまでの距離凡そ400m、標高差凡そ30mの公道に向かい南傾面の立地である。校門から構内3号館の外空間のアプローチ階段下までの距離約230mの点から、3号館敷地まで距離55.2m、標高9.1m。3号館敷地から5号館〜11号館群敷地までの距離凡そ120m、標高差16m。

　3号館から5号館〜11号館群は敷地・外空間共に、傾斜地に構成されている。

　上記のような南傾斜の標高差約25mを有効に使って、これまでの大学のキャンパスに見ることのなかった精神空間を展開することを考えた。

　1977年、木村重雄著『人間にとって芸術とは何か』（新潮選書、1976）を読むうち、宗教ー修験道の項が目に留まり、その中の「石と山岳の信仰」についての記述が参考になるであろうと、心に留めていた。

　その記述とは次のようである。

　「原始社会では、天上界の神が人間界に降臨する時まず山頂にその拠を占めると考えられて、山が神聖化された。

　又農耕が発達すると、山からもたらされる灌漑用水を水神の恵みとする考えが強まった。湯布院町の川西という処に蹴裂権現社があるが、これはこの種の水神を祭る。伝説によると、むかしこの地は満々と水をたたえた大きな湖があったが、神が川西辺りを蹴り裂いたので、湖がだんだんと干上がって今の水田と温泉が出来上がったのだという。そして水落ちとなった。川西の山の端に神を祭り蹴裂権現と称した。このように山岳は神が降臨して人間に恵みを与える神聖であると……。」

　又ブルーノ・タウト著の『日本文化私観』（1880）の一部を抜粋して要約すると、

　「日本の神社は森の中にある。

　森に囲まれて神社がある。

　墓も手入れの行きとどいた、磨きたてられた石ではない。

　むしろ自然に任せて自然の中に溶け込む自然の中に消滅し去っていくもののように見える。

　殊に田舎の墓地は、特定の範囲に限られることはなく、樹の下とか、その他任意の一隅に散在し、全くの自然の風景の中に姿を消してしまい、いわば死者の霊も身体も大自然の中に吸収融合されてしまうようになっている。

　日本の神は自然を超出している神、樹木に囲まれ、自然と共に在る神である。

　日本人は自らの一生を大地から生まれ、やがて大地に帰り、自然の中へ消滅するものとして、少なくとも近代以前は理解していた。」

　工学院大学八王子校舎のキャンパスも日野自動車工業本社"日野台の杜"武蔵野丘陵の一角に位置し、タウトの述べている人間と自然が共有出来る自然の許容限度の中で上手に林を使い、自分達の生活の基礎として半自然性の雑木林を切り拓いて建設されて来た。その自然性の雑木林と共に在った先人は、全く失われた日本人の自然観である日本人の美しい心情を育み、この地方独特の文化を育んだ。

　今回5号館〜11号館・3号館外空間構成展開にあたり、設計テーマの一つの大学として、研究環境として、そこに生活する人びとに潤いを与える空間創りの主旨に沿い人と自然の共存の現代における価値観の変革の問いかけを含めて、その深い意味を漂わせる空間造成と展開を目指した。1989年日野自動車工業本社"日野台の杜"改造主旨によった。

　雑木林は、日本文化の植物的性格を培った一つの姿である。

　現代における価値観とは自然を「消費の対象」としている点を言っている。

　前述の木村重雄氏の著述にある石と山岳の信仰については、筆者がモチーフとしている。

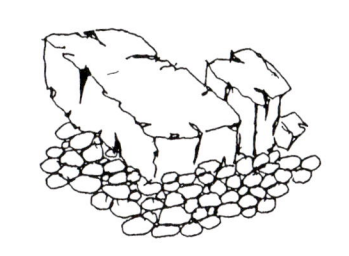

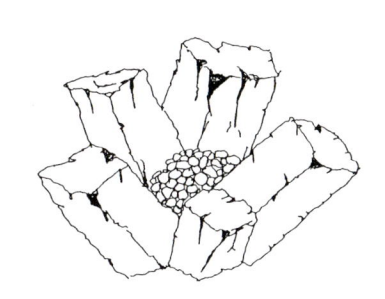

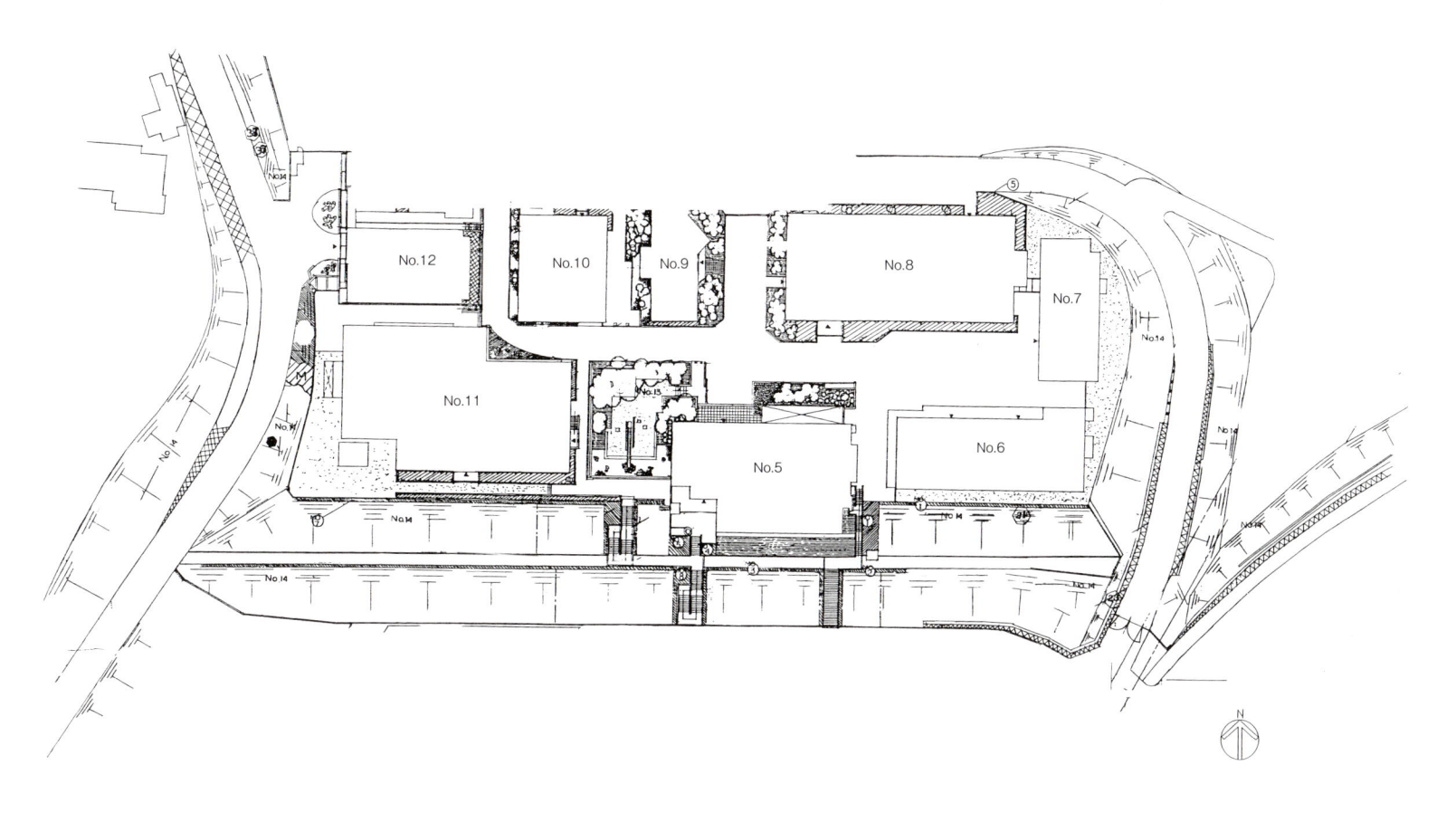

1986年
用石の種類
石　積　　　　　新小松石間知石、芦野石切石
見切石　　　　　稲田石切石
床鼻石　　　　　　〃
階段石　　　　　　〃
床貼　　　　　　稲田石板石
舗　床　　　　　塩山石ピンコロ、福島石ピンコロ
立石、石組　　　芦野石雑石
単　体　　　　　新小松石
ベンチ　　　　　稲田石
看板石　　　　　イタリア産みかげ石
方位石　　　　　稲田石
館名石　　　　　　〃
コンクリート打放し　塗装仕上、ハツリ仕上
舗　床　　　　　塩山石ピンコロ、アスコン、アサノ洗出平板
水の様態　　　　流れ、落水

※詳細図はディテール図面編の176〜177P参照

Kogakuin University Hachioji Campus, Bldgs. No.5 to 11 *'Gai-Kukan'* 1986

The *'Gai-Kukan'* for surrounding Bldgs. No.5-No.11 completed in 1986, *'Gai-Kukan'* for the Bldg. No.3 was completed in 1987, both of after the building itself finished.

The distance is about 400m between the gate, facing road and the height of the Bldgs. No.5-No.11, stand. The elevation of the height is about 30m and its facing south. Distance between the gate and approach of the *'Gai-Kukan'* for Bldg. No.3 is 55.2m, and the elevation is 9.1m. The distance between the sections of Bldg. No.3 and the surrounding Bldgs. No.5-No.11 is 120m and the elevation is 16m.

The whole area is on a slope, including the *'Gai-Kukan'* for Bldg. No.3 and the surrounding Bldgs. No.5-No.11.

I thought develop a spiritually influencing space in university campuses never seen before by take an effective way for the elevation of some 25m.

In this case, I had inspiration when I reading the book *What is Art for people* by Shigeo Kimura (Shincho Sensho, 1976), I was attracted by the section of religion-training, I thought it would be use the inspiration of which this description "the belief on stone and mountain" some day.

The description is as follows; In primitive societies, it was say, the god in heaven one stay a while on top mountains when comes down to the human society before, caused mountains was considered sacred place.

As farming developed, people thought that water for irrigation from the mountain was blessing by the god of water.

There is temple of Kesaki Gongensha in Kawanishi district in Yufuin town, is enshrined such god of water. According to a legend. In old times there was a lake full of water, but one day god kicked and opened the bank around Kawanishi, so water gradually flew out and it left rice field and hot spring as we see today. The place became Mizuochi today.

People built a shrine at the foot of a mountain in Kawanishi and called it Kesaki Gongen, meaning kick-rupture-god. In this way mountains are where god comes down and blessing us and therefore are sacred.

Bruno Taut says in his book *A Personal Observation on Japanese Culture* (1880) excerpted and summaried as follows; 'Japanese shrines are built in woodlands, and the graveyards in the woodlands are not made up of well-kept or polished tombstones; As if they were just left it to nature, being merge into nature gradually, and disappear last. Especially the graveyards in the countryside are not limited to any designated area, but it can be found under some trees or in other unspecified spaces. It was totally lost from viewing in nature, in a sense as if the body and soul was in harmony with the great natures. Japanese gods be transcend to nature, surrounded by trees and also coexist with nature. Japanese recognized the death as which be born from the earth and go back to the earth, and as disappearing into nature, at last before present day.'

Hachioji Campus of Kogakuin University and 'Forest of Hinodai' of Hino Motors, Company Main Office on Musashino-Hills area.

It was built on the woods, which was logging but yet within an allowable limit as Taut said like that, people could live with nature in harmony.

The ancestors lived with the truly natural woods here, have fostered the beautiful Japanese mind, and developed a unique culture what the affection toward nature.

It's aims of the design of the *'Gai-Kukan'* for Bldgs. No.5-No.11 and No.3, was to create natural environment in which relaxation necessary for students, teachers, and people who live in around.

This *'Gai-Kukan'* is intended to recognize difference that living with natures and their modern sense of value, to people. By modern sense of value, I mean people nowadays regard nature as an object of consumption.

The design principle follows that of the renovation design of 'Forest of Hinodai' of Hino Motors, Company Main Office, built in 1989, and it is supposed to give people philosophical suggestions.

I thought woods are representation of the botanical nature of Japanese culture.

I have set a motif by my belief based of stone and mountain worships.

1986
List of stones
Stone walls: Shinkomatsu-ishi, Kenchi-rocking-wall. Ashino-ishi, cut.
Mikiri-ishi: Inada-ishi, cut.
Tokobana-ishi: same as above.
Stone treads: same as above.
Floor finish: Inada-ishi, flag stone.
Paving: Enzan-ishi, granite setts. Fukushima-ishi, granite setts.
Standing stone, Stone work: Ashino-ishi, fragments.
Single stone: Shinkomatsu-ishi.
Bench: Inada-ishi.
Signboard: Italian granite.
Direction indicator: Inada-ishi.
House name plate stone: same as above.
Cast concrete: Painted, crack-shaving-finish.
Paving: Enzan-ishi, granite setts. askon. Asano washed plate.
Water features: Stream, waterfall.

※ details 176-177p

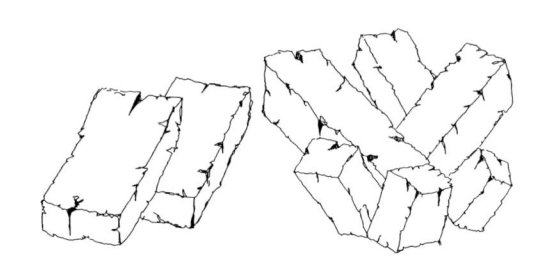

115

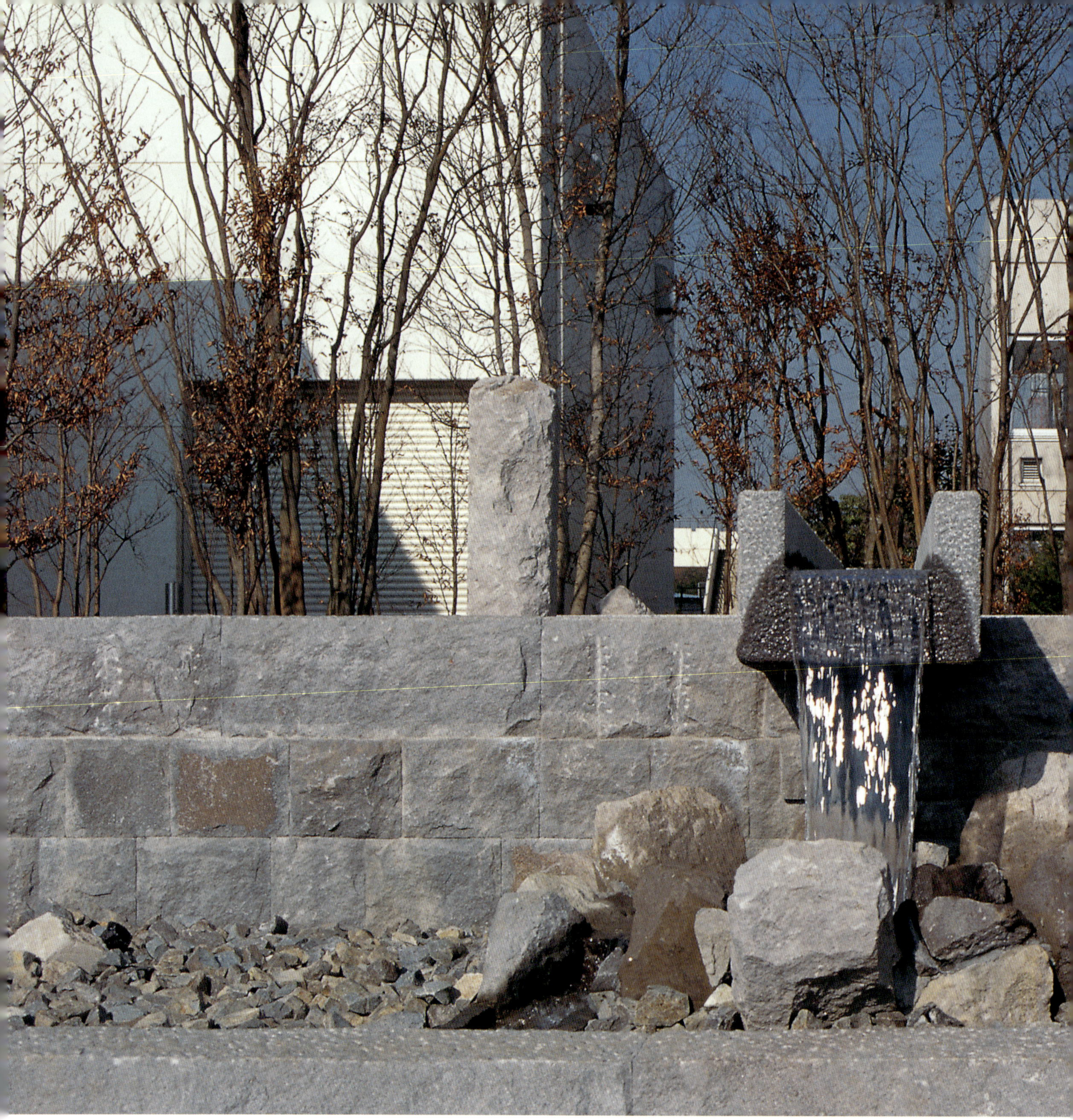

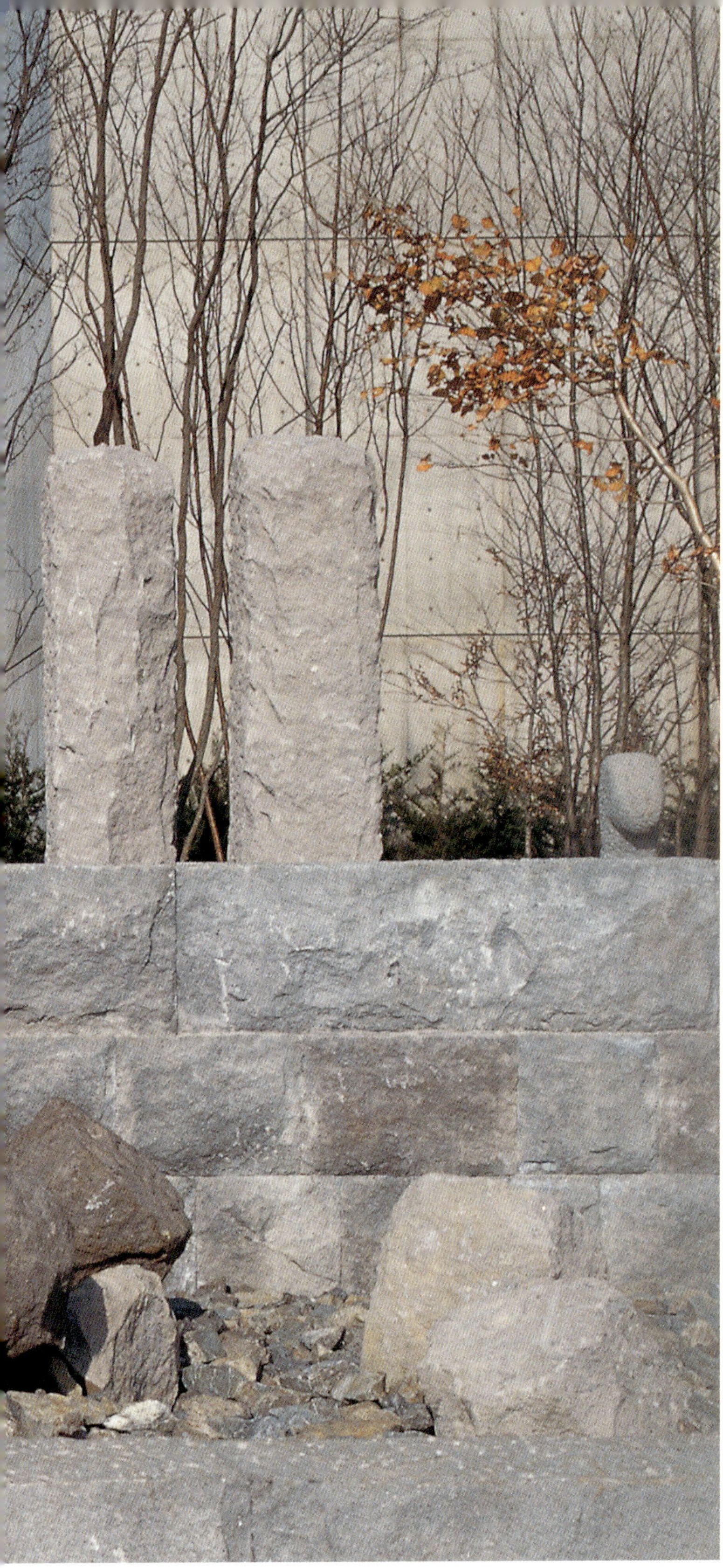

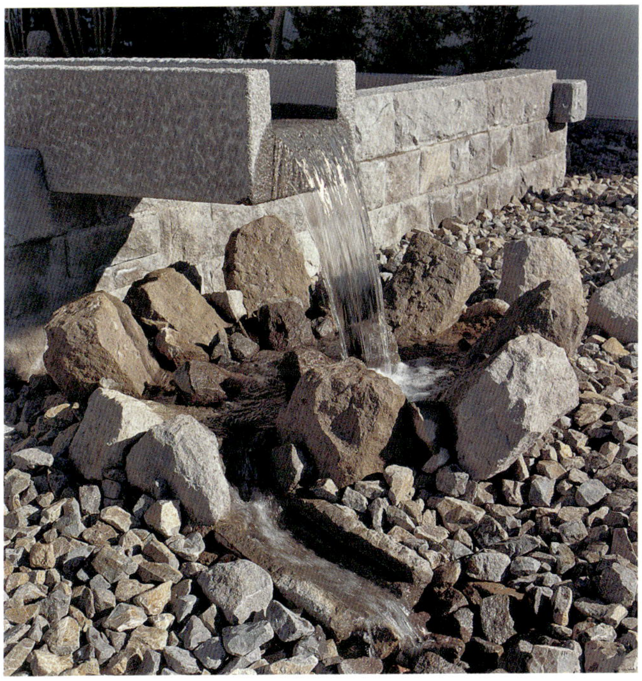

工学院大学八王子校舎 3 号館 外空間 1987
Kogakuin University Hachioji Campus, Bldg. No.3 *'Gai-Kukan'* 1987

119

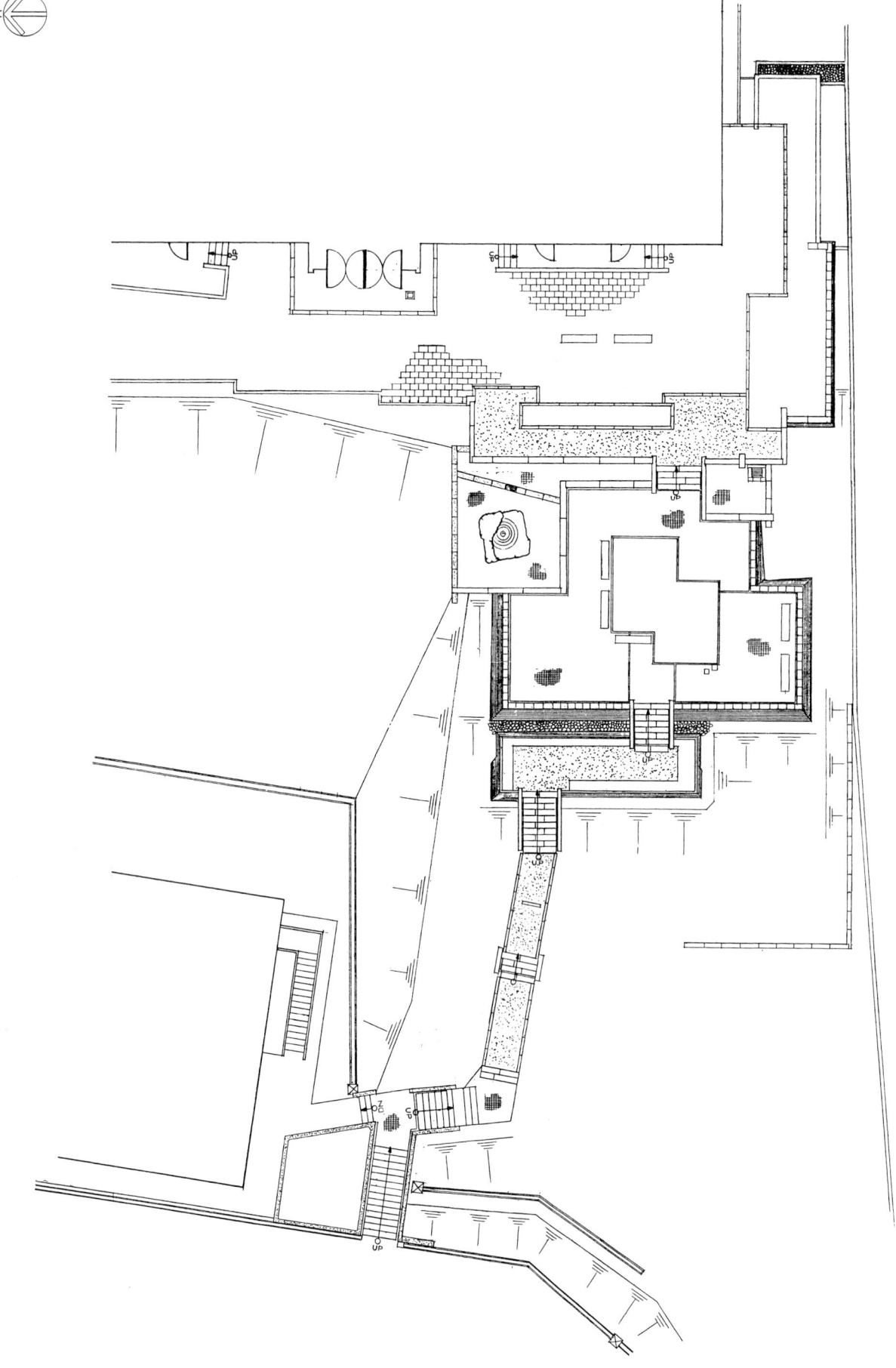

平面図 1/300

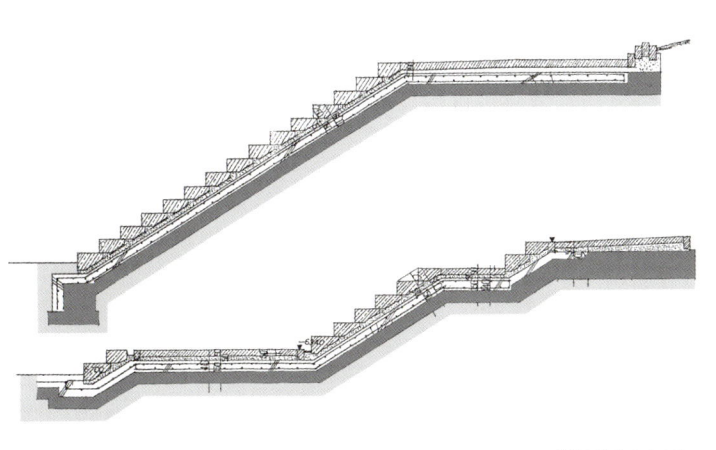

断面図 1/100

121

工学院大学八王子校舎3号館 外空間 1987

Kogakuin University Hachioji Campus, Bldg. No.3 'Gai-Kukan' 1987

　この空間は、先の5号館〜11号館群の小広場に語りをつなげた。この空間も5号館群外空間同様、斜面を登りつめた処に3号館が位置するということから、3号館に至る起点となる約9m余りの高低差を持つ駐車場より3号館に至る間のこの外空間は、途中に小広場を設けるほかは、石積・石の階段・点景として立石の石組などの要素でまとめた。

　この外空間には、積み上げ方式を表す新小松石間知石の石積、立っている精神と抵抗の精神を表す芦野石の立石の石組、人間の刻印として水の神を表す石の水盤、そして頂の3号館校舎への道の導きとしての稲田石（花崗岩）が素材の心の階段、以上のそれぞれを、幾何学を思わせるように展開した。果たしてこんな美旬が、表れているだろうか。

　私の願いは、この空間が、永い厳しい風雪の時を経て、廃墟の美として姿を現す日まで遺ってほしいといった秘かな願いである。

　完成して私は心の裡で、工学院大学八王子校舎5号館〜11号館小広場－3号館外空間－日野自動車工業本社「日野台の杜」（外空間）が一つの空間の物語として完結したことを喜んだ。そして、1971年から87年まで16年掛かりの道程の中で生まれて往った幾つかの心に遺る空間、そして1991年に完成した2つの公共空間まで、ともすれば挫折へ傾く私の内面を支えてきてくれた見えざる超越者に感謝すると同時に、生まれた作品の一つ一つは超越者の作品ではないのかと迷う私である。

1987年
用石の種類
石　積　　　　　　　　　新小松石間知石
　〃　　　　　　　　　　コンクリート間知石
コンクリート擁壁笠石　　　新小松石切石、稲田石切石
コンクリート間知石積一部笠石　芦野石切石
縁　石　　　　　　　　　中国産みかげ石、芦野石切石
階段石　　　　　　　　　稲田石
階段袖石　　　　　　　　　〃
階段石　　　　　　　　　一部花崗岩、大東石
床鼻石　　　　　　　　　稲田石
ベンチ　　　　　　　　　　〃
水の様態　　　　　　　　湧水、流れ

※詳細図はディテール図面編の178〜181P参照

This 'Gai-Kukan' is connected with the small plaza of the group of Bldgs. No.5-No.11, like a one of the group. Bldg. No.3 placed upside on slope as the same.
Its location was that, height from attention point as the parking lot to Bldg. No.3 is more than 9m, include a small plaza in the middle of the way.
And elements are, stone walls, stone steps and structures of standing stones as object.
The mainly materials are; stone walls of Shinkomatsu-ishi, Kenchi-rocking-wall, standing stones of Ashino-ishi representing standing spirit and resistance spirit, stone water basin representing god of water as reflected human, and stone treads as mind stairs, made of Inada-ishi, be onductor to Bldg. No.3 on top of the hill. These elements were arranged in a way to remind the viewer of geometry.
Would it really be expressed such a worthy artistic descriptions? My furtive desire is that this space will last through the hardship of all kinds of weather and stand some day as a beautiful remain.
After completion, I am very pleased to be appear as one of continuous story that the three 'Gai-Kukan', it's for Bldgs. No.5-11, for Bldg. No.3 in Kogakuin University Hachioji Campus, and for Hino Motors, Company Main Office, 'Forest of Hinodai'. Many 'Gai-Kukan' were born during the for 16 years from 1971 to 1987 and they still live in my mind with the two public spaces completed in 1991.
I would like to appreciate for help to the invisible, transcendental entity who kept me highly motivation, despite I succumb to my discouragement.
I really wonder if all I produced is what work of transcendental entity.

1987
List of stones
Stone walls: Shinkomatsu-ishi, Kenchi-rocking-wall.
Stone walls: Concrete, Kenchi-rocking-wall.
Topping of concrete wall: Shinkomatsu-ishi, cut. Inada-ishi, cut.
Concrete, Kenchi-rocking-wall, Part of topping stone: Ashino-ishi, cut.
Stone curb: Chinese granite. Ashino-ishi, cut.
Stone treads: Inada-ishi.
Stone piers for steps: same as above.
Stone treads: Part of granite. Daitoseki.
Tokobana-ishi: Inada-ishi.
Bench: same as above.
Water features: Spring water, stream.

※ details 178-181p

立面図 1/200

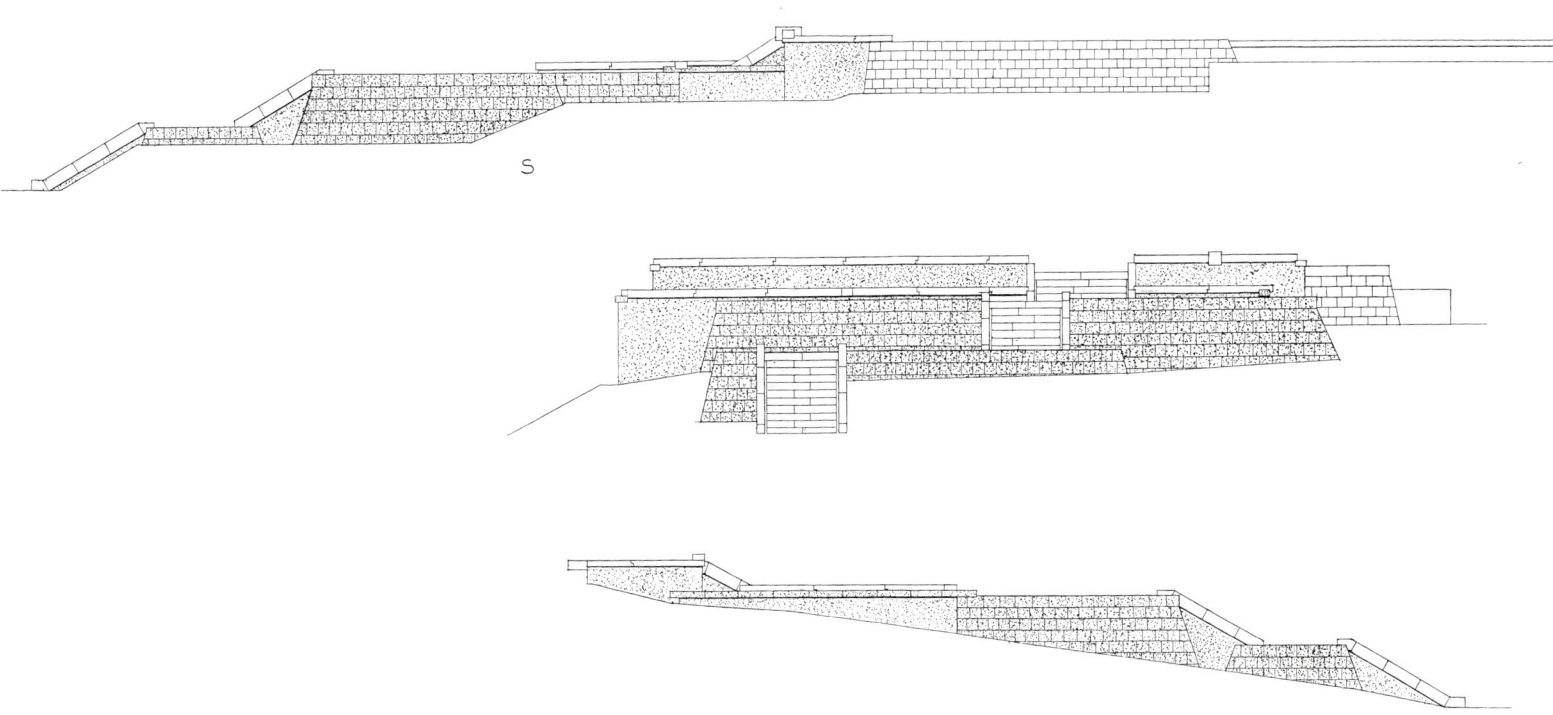

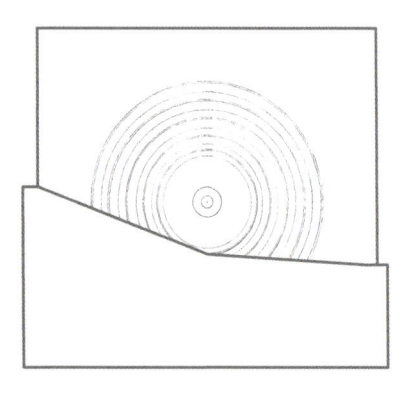

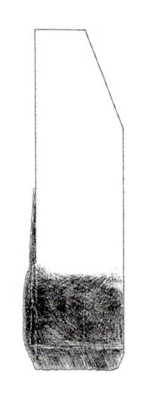

湧水単体平面図 1/50

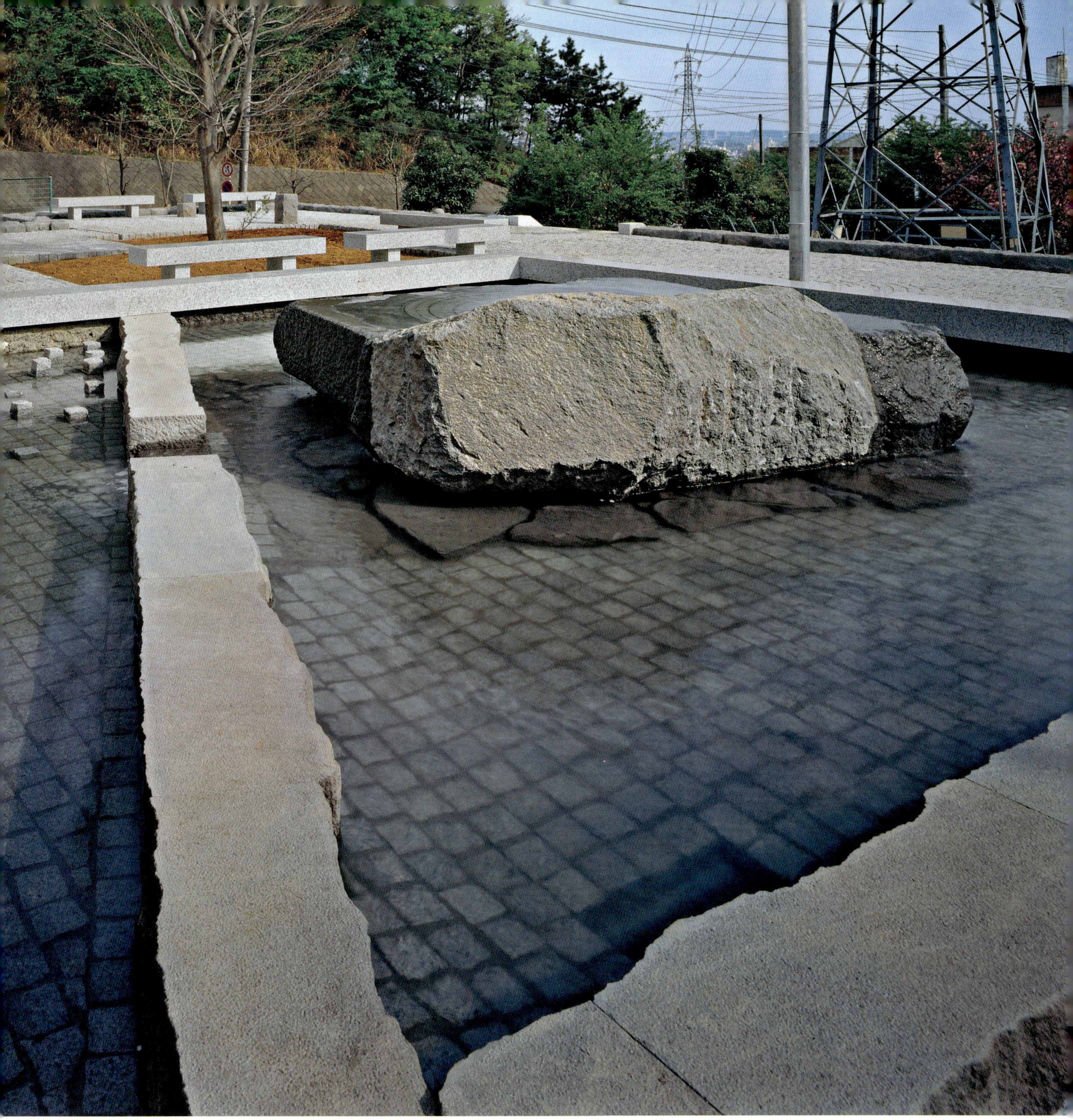

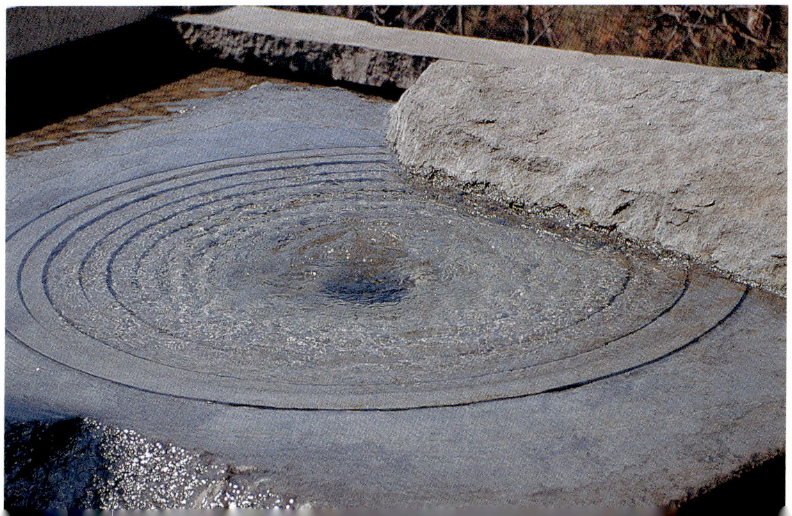

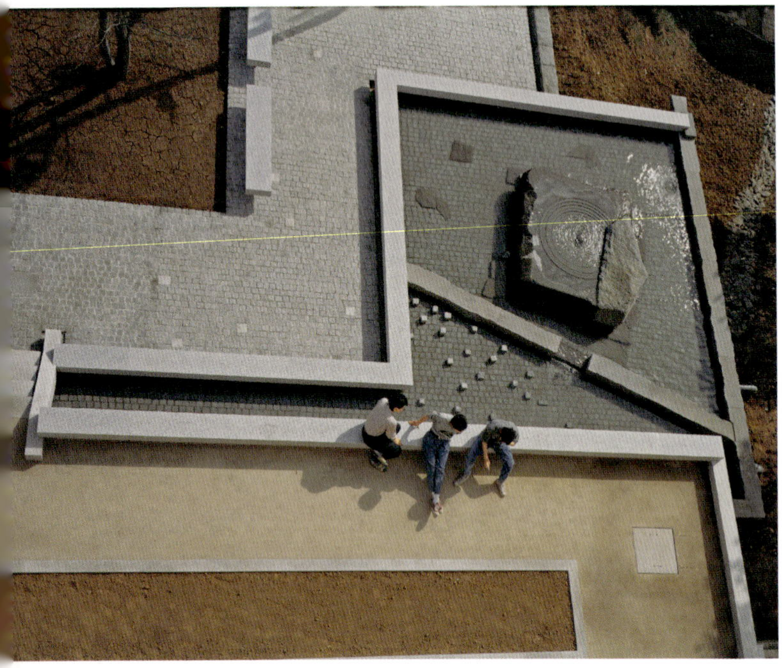

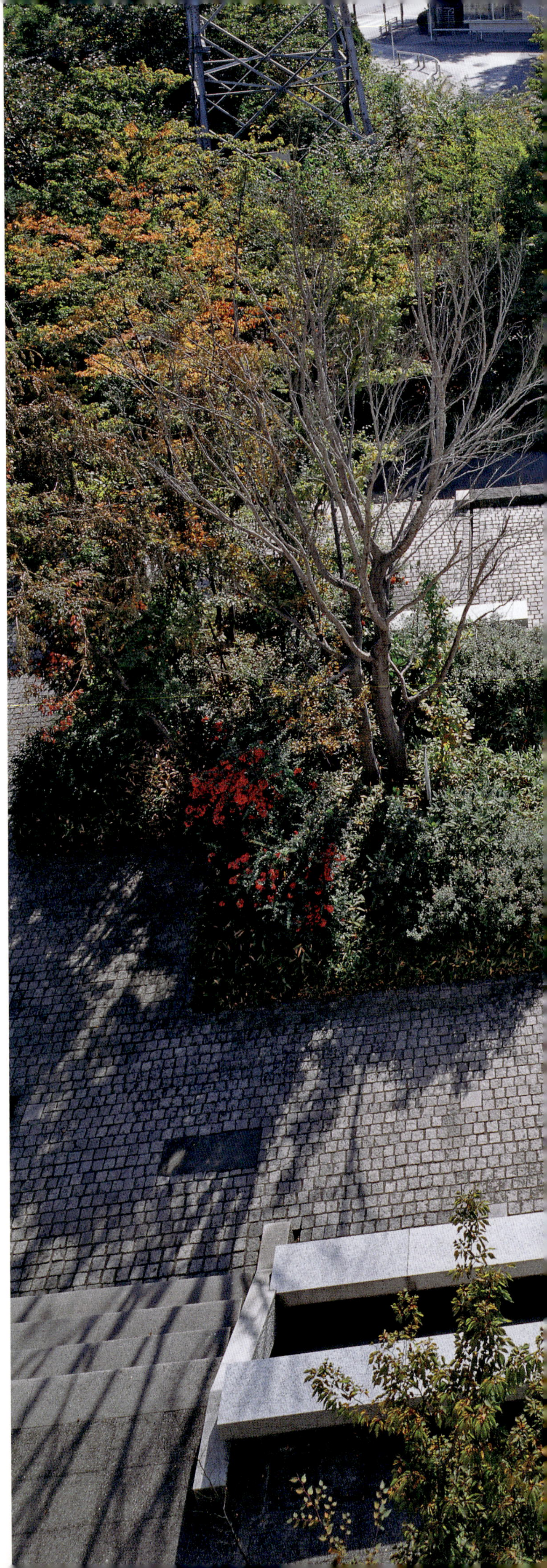

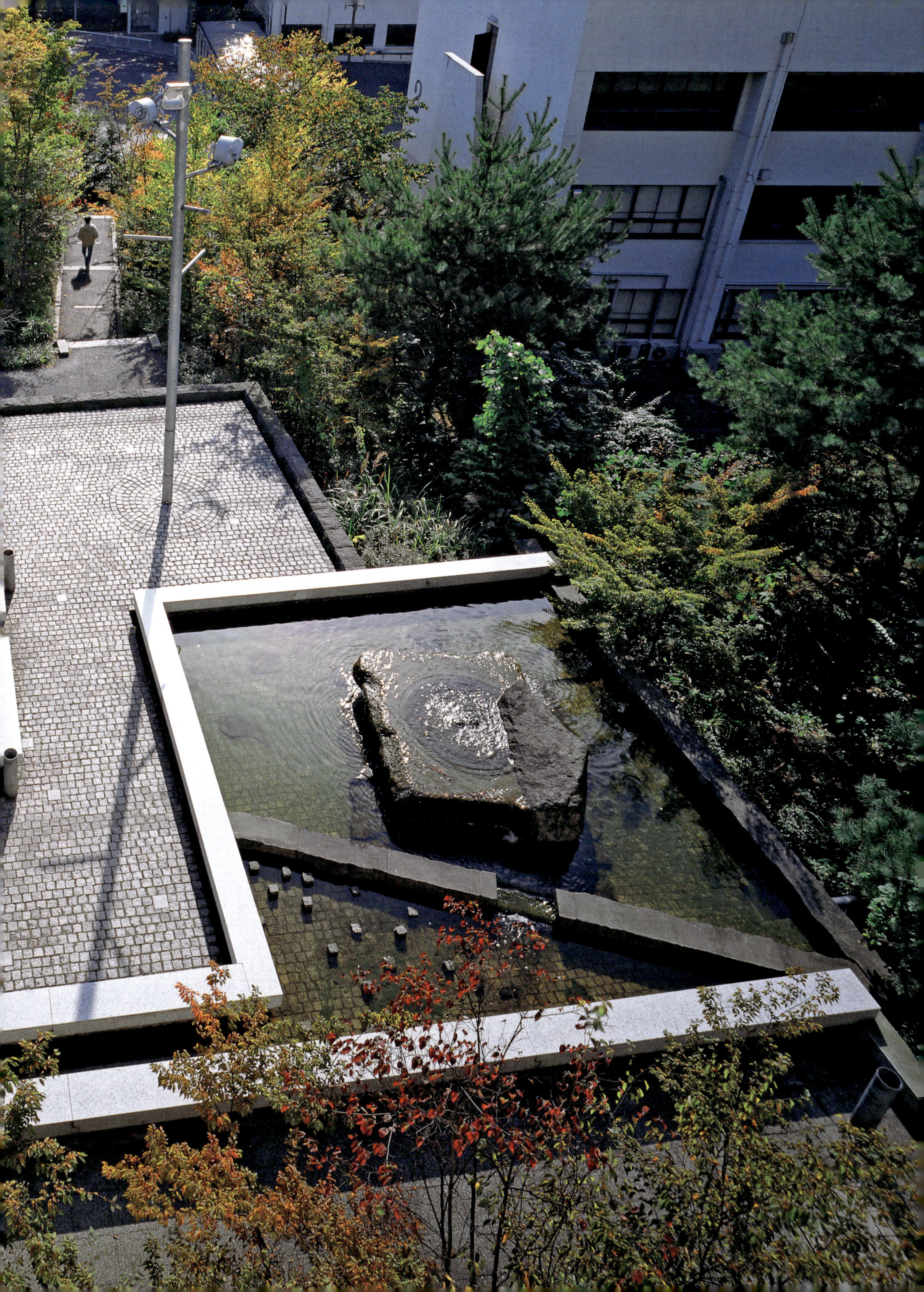

スコープ本社 GLASS HOUSE 外空間 1991
SCOPE Main Office GLASS HOUSE 'Gai-Kukan' 1991

1/200

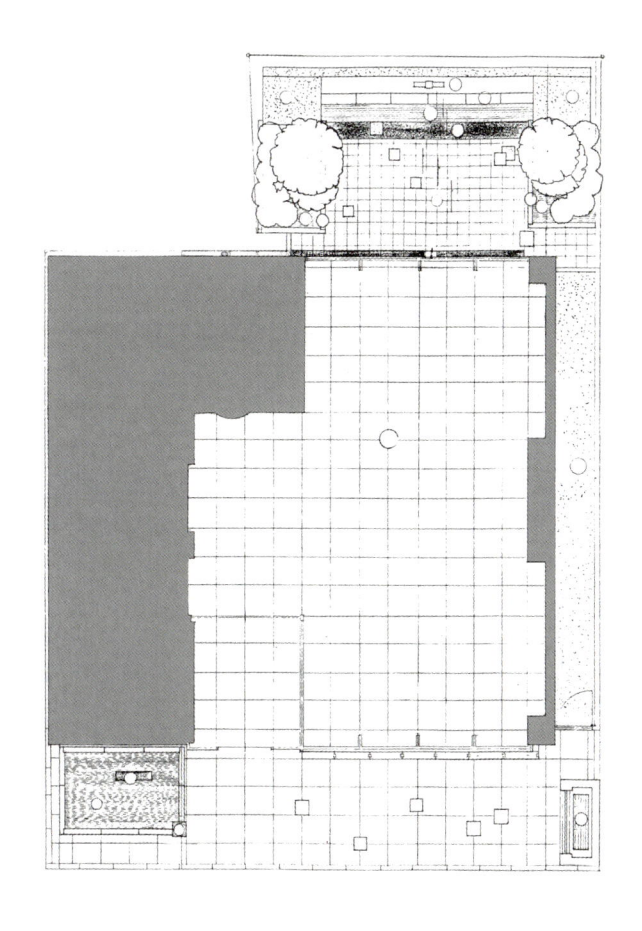

129

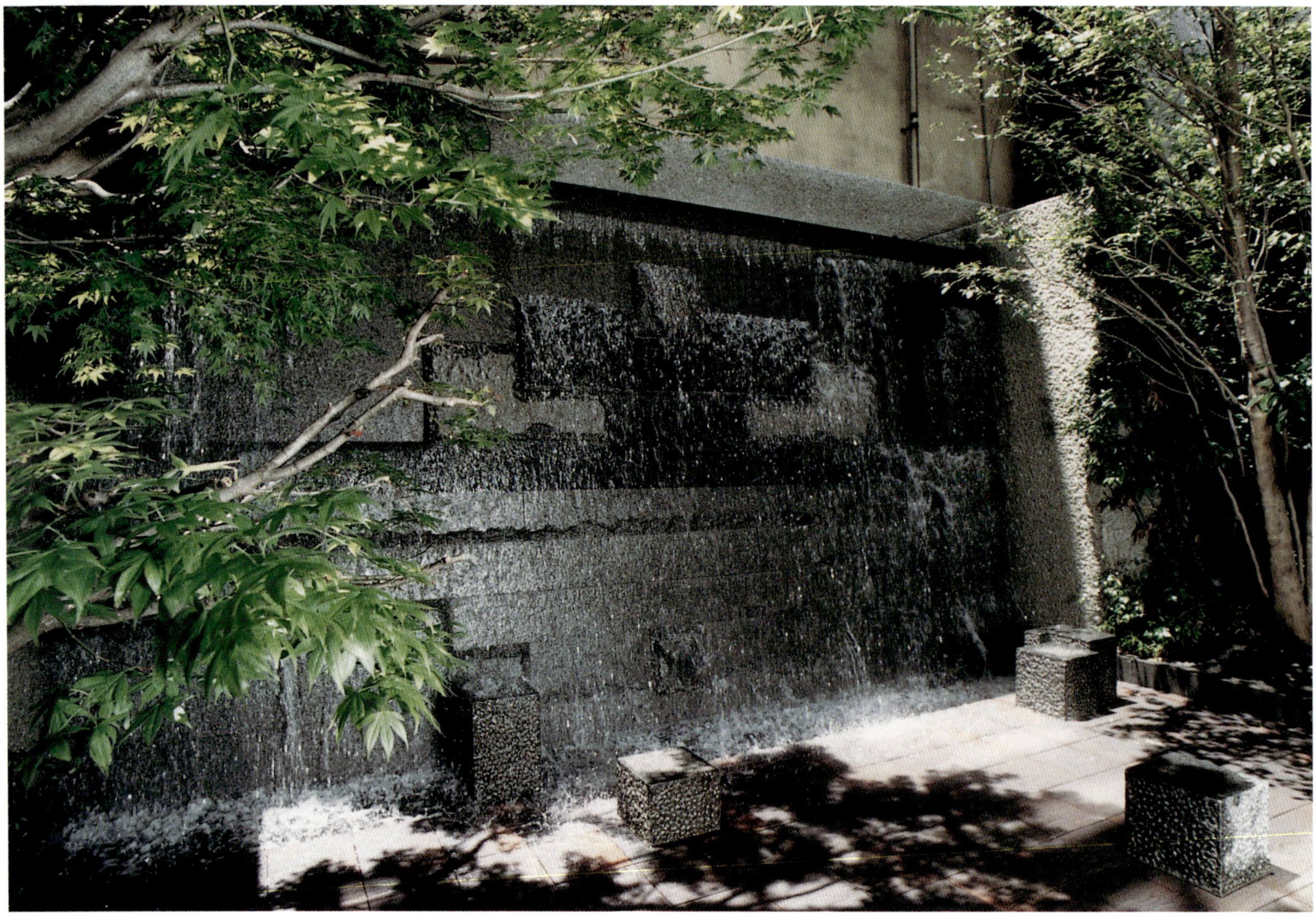

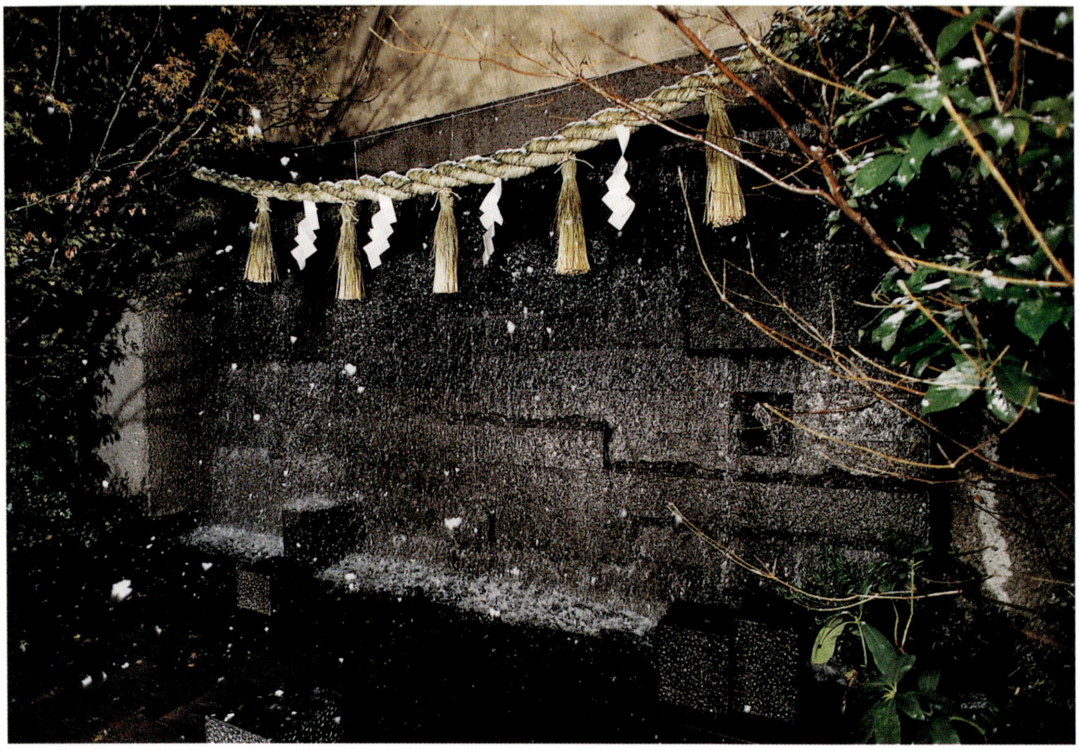

スコープ本社 GLASS HOUSE 外空間 1991

　都営地下鉄東西線神楽坂駅下車、近くの赤城元町の赤城神社西脇の坂を下りしばらくして最初の十字路に至る。これを左に廻りしばらく歩むと、「GLASS HOUSE」の前に至る。この界隈が新宿区築地町9番地で、GLASS HOUSEから神田川筋に至る道筋の西五軒町、水道町を過ぎたあたりで目白通りに至る。西五軒町と目白通りの角辺りに出版社雄鶏社があり、水道町と目白通り角辺りには印刷の三晃印刷がある。西五軒町も水道町も印刷・製本・出版を業とする中小の企業が軒を並べひしめき、隣接する築地町の界隈も多聞にもれず、同じような企業がひしめく猥雑な眺めの日々が続く処である。その上、この界隈には車優先かといぶかられるような"そこのけそこのけお馬（車）が通る"が常識のような道路が縦2本、横2本程通るのみで、非情さ溢れるとしか言い様がない。

　スコープ本社 GLASS HOUSE は、このような雰囲気の車優先の公道との境界からファサードまで約3m余りを公共に開放して、図のような敷地に立つ6階建ての建物である。

　社長の横山寛氏の依頼内容は、九段下のホテルグランドパレスの滝のある雰囲気が気に入っているので、どうだろうというものであった。

　検討の結果、公道と一体化されたファサードとこれに続くエントランスホールを通し、滝のある空間まで視覚的に一体化可能の空間構成とすることに決めた。表現を変えると、滝のある空間と建物とを背景に道と一体化し、人びとを快適に抱き込む空間構成とした、ということであろう。

　この空間は、1991年3月に完成した。

　完成した空間を眺めて、喜びの言葉をと言われれば、私は次のような表し方をするだろう。

　GLASS HOUSE の外空間は小規模ながら、個々の敷地や建物を背景として道と一体化し、新しい形と人との関わり合いを持ち込んだ空間として、これまでの外空間とは趣の異なったものとして、大きな成功をおさめたように思われる。

1991年
用石の種類
石　積　　　　　黒みかげ石ラステンバーグ
縁　石　　　　　〃
転　石　　　　　〃
床　　　　　　　志野陶石
石積附属壁　　　コンクリート打放しツツキ仕上
館名石　　　　　ラステンバーグ
キャビネット扉　方形型鋼
砂利など　　　　那智黒、石灰砕石

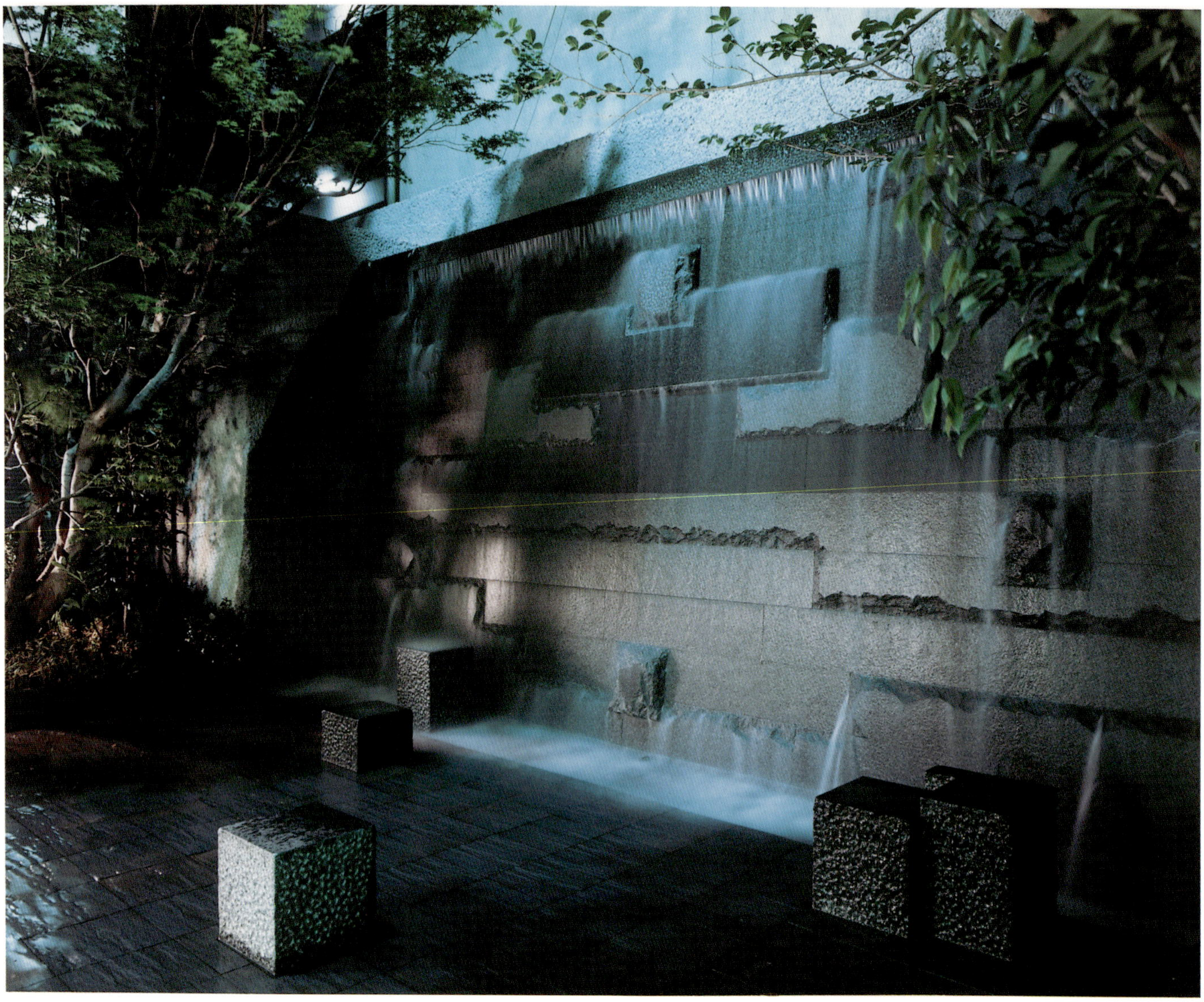

SCOPE Main Office GLASS HOUSE '*Gai-Kukan*' 1991

Near the Kagurazaka Station on the Tozai Subway line, walk down the slope to the west of Akagi shrine in Akagimoto-machi and turn left at the first intersection, after walking for a while there will be a building called 'GLASS HOUSE' on the left. This neighborhood is ,9 Tsukiji-cho, Shinjuku, and the road continues on to Mejiro-dori after going past the 'GLASS HOUSE', and through Nishi Goken-cho and Suido-cho to the Kanda River. The publishing company Ondorisha is at the corner of Nishi Goken-cho and Mejiro-dori, and the printing firm Sanko Insatsu is at the corner of Suido-cho and Mejiro-dori. The neighborhood has been cluttered, lately by such small, printing, binding and, publishing companies. The more this kind of development takes place the uglier the area becomes, and it extends to the adjacent neighborhoods like Tsukiji-cho. To make matters worse, in this area there are two streets busy with vehicular traffic crossing lengthwise and crosswise which take precedence over all else. It is like in the old days when the Daimyo's horse would pass with shouts of 'Stand aside. A horse is passing.' Utter callousness is the only description.

The GLASS HOUSE is a 6 storey building set back from the street by a 3 meter narrow public space. The president of the company, Hiroshi Yokoyama, mentioned that he liked the atmosphere of the waterfall at the Hotel Grand Palace in Kudanshita.

After careful consideration, I decided to create a space where the building façade, entrance hall, and public street were all visually unified with the waterfall. In other words, the spatial composition is designed to unify the space with the waterfall, the building, and the street that has them 'at the background, comfortably embracing the people'.

This space was completed in March 1991.

If I were asked to comment on the completed space, I would have to say that I feel it is a great success in that it has a completely different atmosphere from the other spaces I have created to date. Though small, 'the space uses the ground plane and building as an effective backdrop that is unified with a road'. It creates an interaction between these new forms and the people who use them.

1991
List of stones
Stone walls: Black granite 'Rustenburg'.
Stone curb: same as above.
River cobble: same as above.
Paving: Shinotoseki.
The wall adjacent to the stone walls: Cast concrete with finished with prodding.
House name plate stone: 'Rustenburg'.
Door of cabinet: Rectangular shaped steel.
Gravel: Nachiguro. Cracked limestone.

三番町 KS ビル 外空間 1991
Sanban-cho KS Bldg. *'Gai-Kukan'* 1991

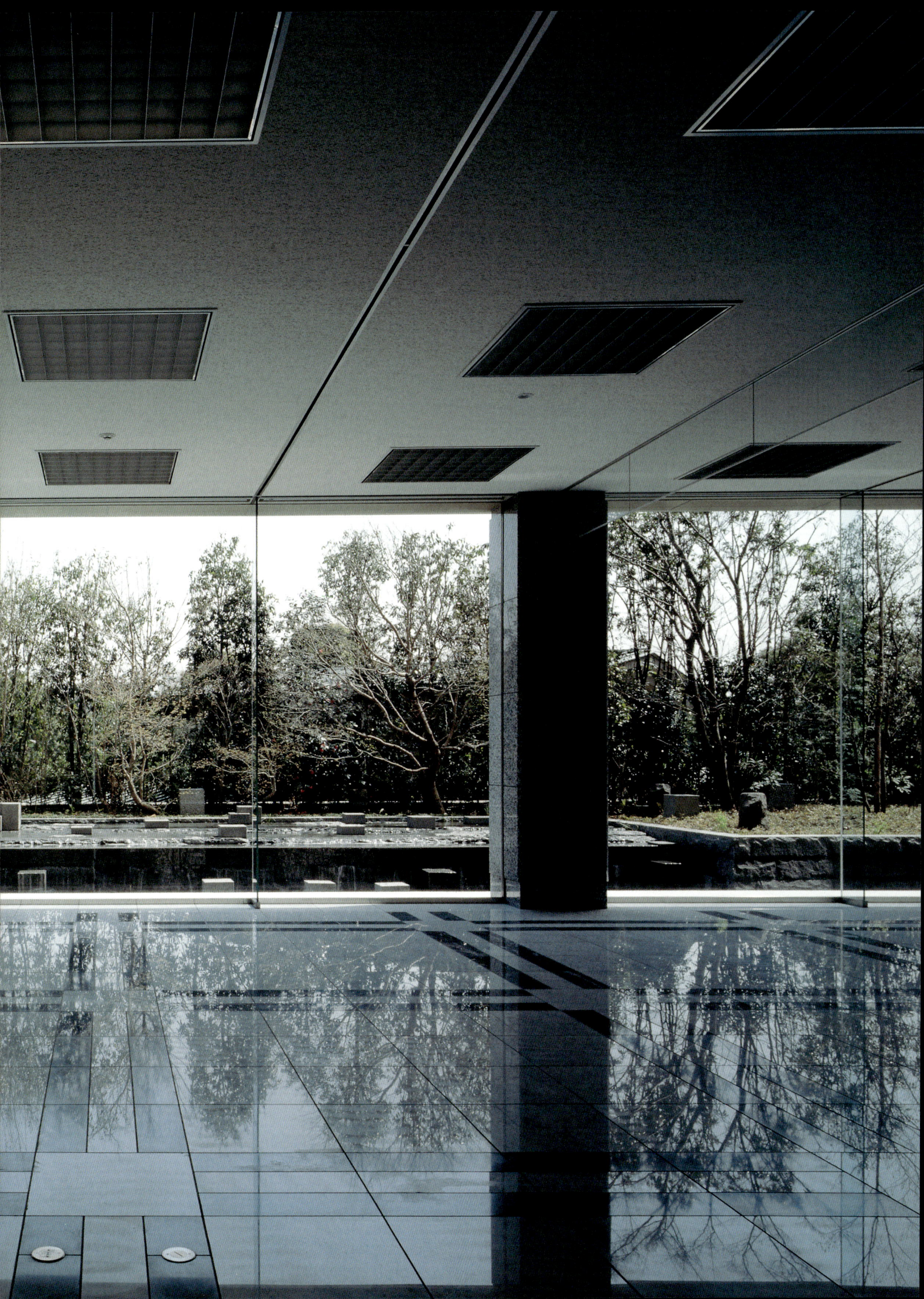

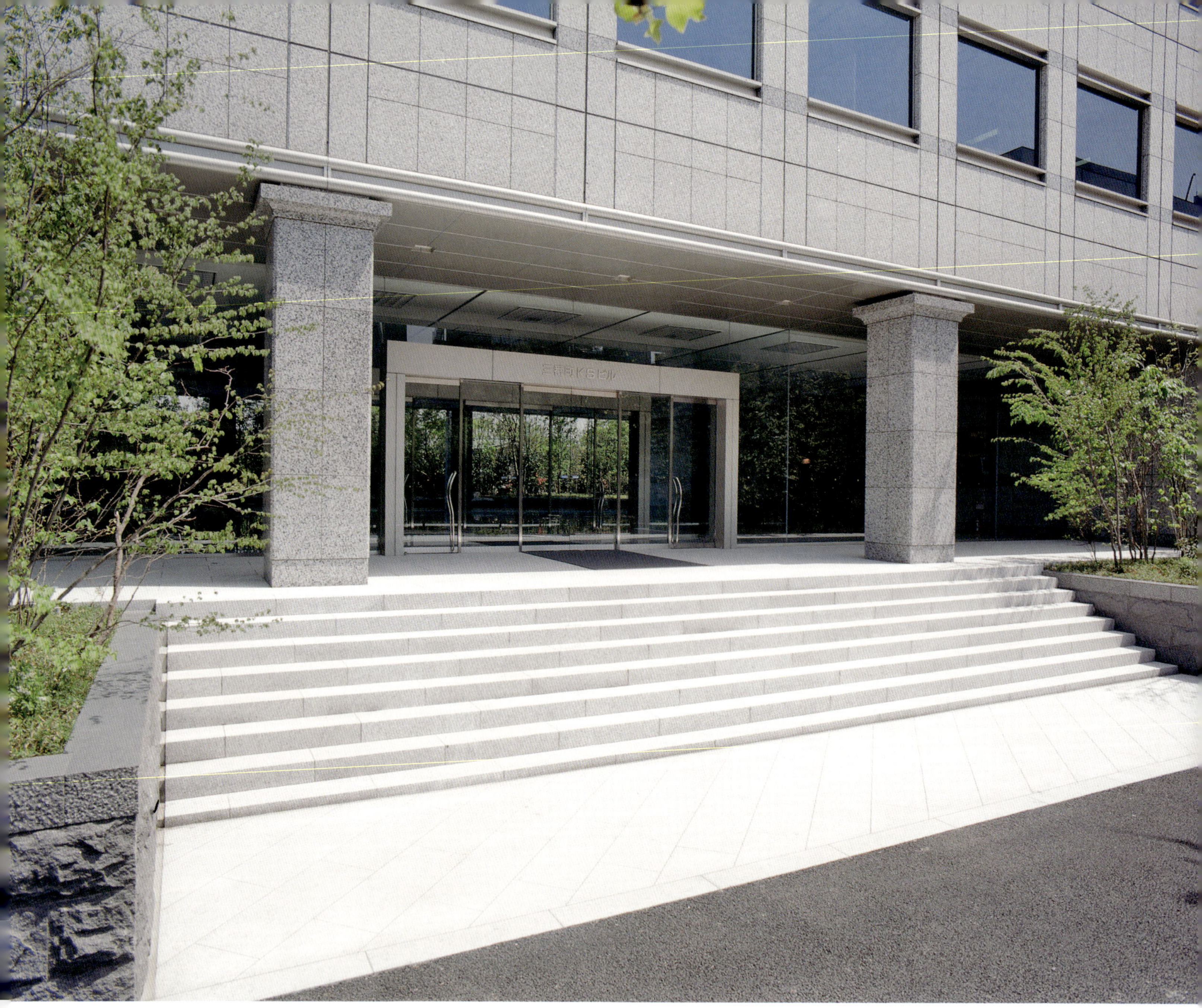

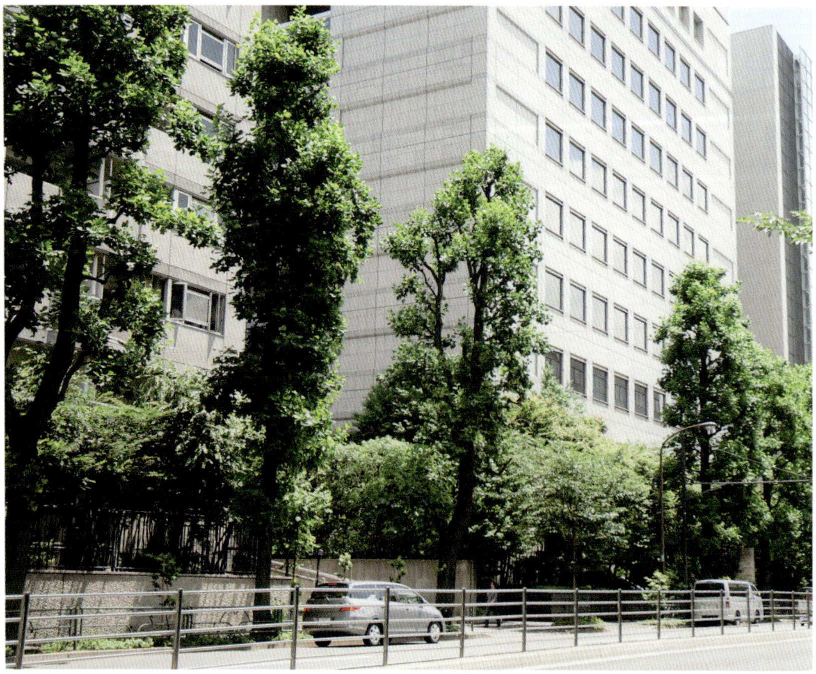

137

平面図 1/400

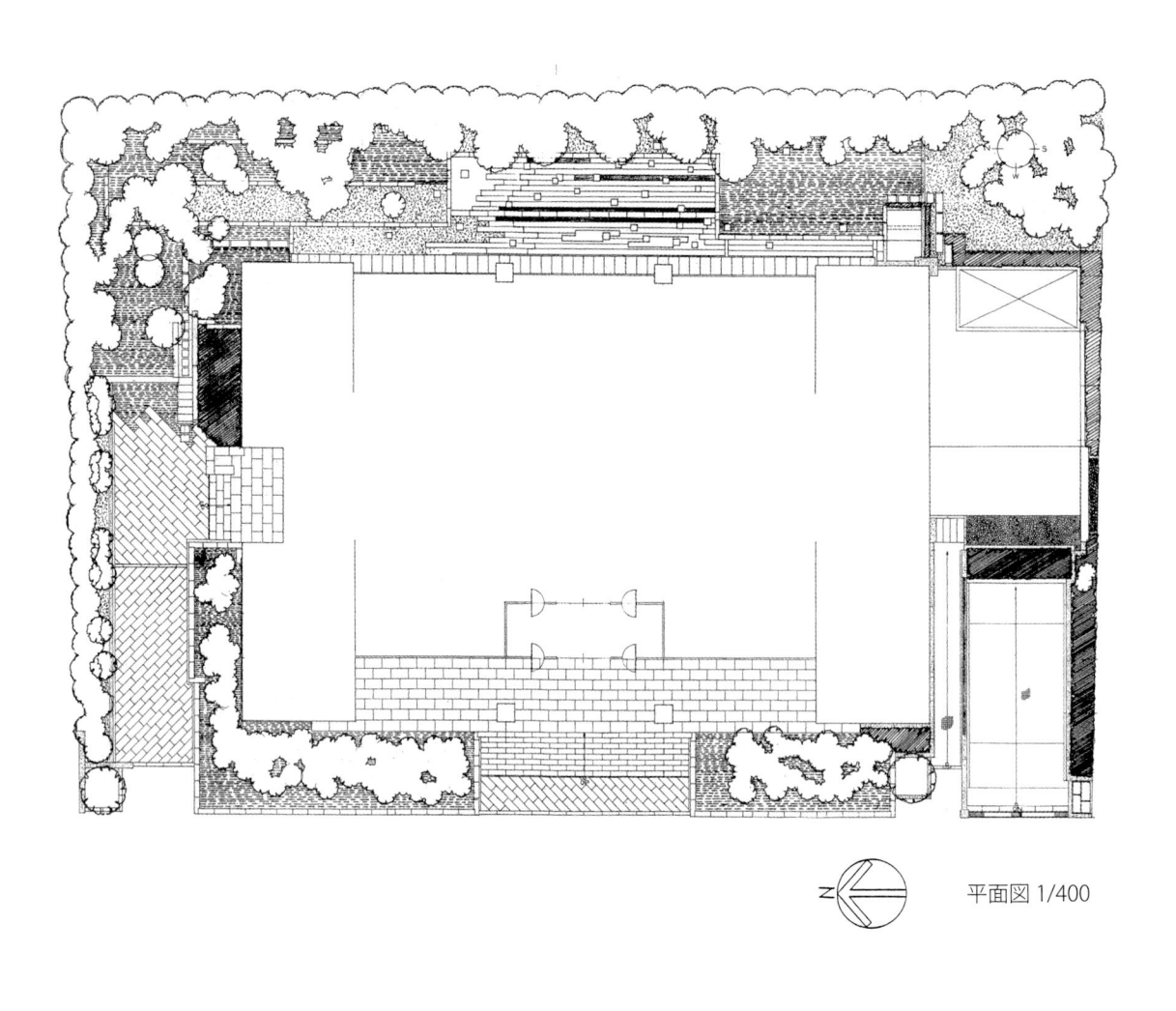

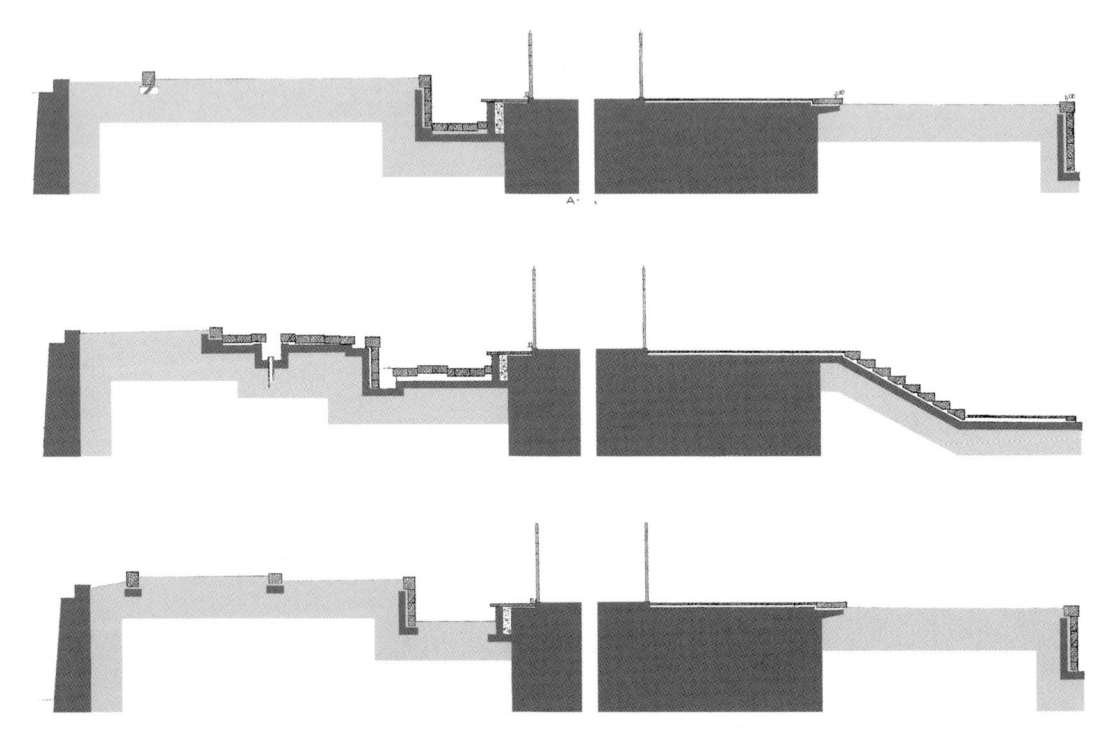

断面図 1/150

138

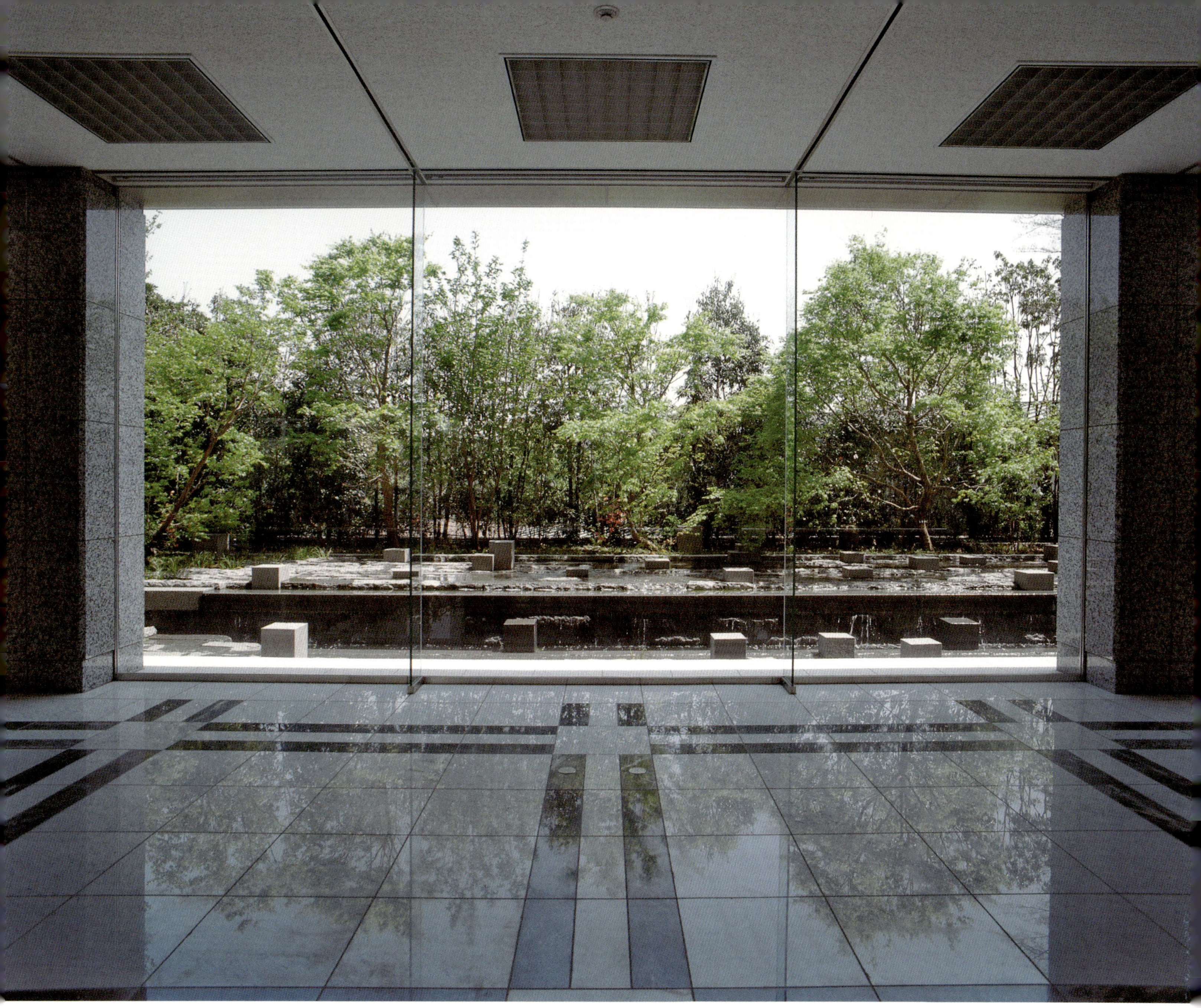

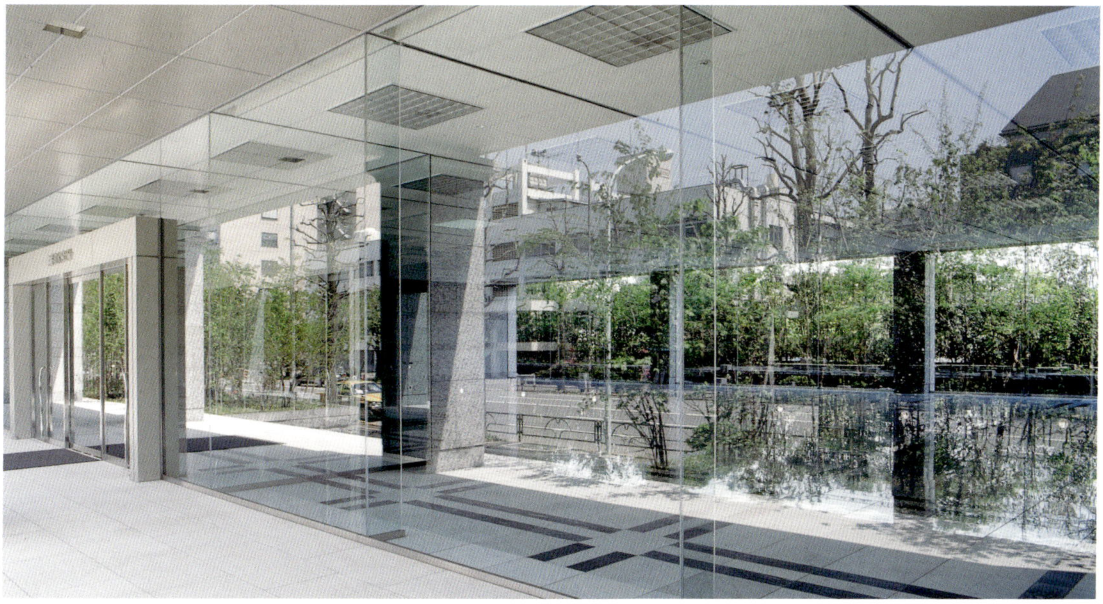

三番町 KS ビル 外空間 1991

　皇居外苑を取り巻く濠と石垣、外桜田門から三宅坂・半蔵門・千鳥ケ淵にかけて、高い芝土居の上の石垣と老松、対岸に近代ビルが立ち並んで共に濠の水面に映えている風景は、今の東京の中で最も美しい景観のように思う。

　靖国通りを九段坂下から新宿方向へ進み、九段坂上のインド大使館角を左折、内堀通りに入り三宅坂方向へ至る途中の二松学舎大学附属高校を過ぎた辺りが、このビルが建つ三番町で、千鳥ケ淵と内堀通りに挟まれて、千鳥ケ淵側にフェヤーモントホテル、宮内庁用地、戦没者墓苑等が並び、内堀通り側は宮内庁用地、三番町 KS ビル、相賀邸と道路をへだてて飛島建設本社が建つ、緑の多い閑静な雰囲気を持つ処である。一例を掲げれば、千鳥ケ淵墓苑として人口に膾炙している緑多い戦没者墓苑。他方、桜の名所ともなっている千鳥ケ淵の土手などのように、この界隈は、三番町の中でも最も緑の多い処である。

　「三番町 KS ビル」は、このような雰囲気を持った地に図のような配置で建つ、外装が辺りの緑によくマッチする韓国産みかげ石・木花石による 10 階建てのビルで、この界隈では目新しいデザインである。その上喜ばしいことは、このビルの外部空間が道と一体化された空間、即ち道がふくれてまわりの建物のファサードを効果的な背景として人びとを快適に抱き込むような空間に近付き得られるような素地があったことである。内心、これはこれまでにない快適な空間を生み出せると喜んだ。その上この地には、20 数年前の九段坂下のホテルグランドパレスへ外空間の計画の下見に訪れた折と同様の霊気が漂っていた。これまでこの気配を感じた空間では、不思議と空間の展開がスムーズに流れ、出来上がった空間は迫力あるものとなった。その迫力ある空間は、いつも人間の生活と深く関わり合った遺跡の匂いをその外空間を訪れる人びと、その近辺に介在する人びとに感じさせるといわれる。私は、迫力ある空間とは、そのような物と人とを結ぶ意味を人の手で創り出すことではないかと考えている。

　これまで創り出した空間で、遺跡の匂いを漂わせているといわれる空間の代表的なものは、

京王プラザホテル 4 号街路　外空間　1971 年
日野自動車工業本社「日野台の杜」　1972 〜 90 年
ホテルグランドパレス　外空間　1972 年
東京銀行青山寮　外空間　1975 年
サンかがや　外空間　1976 年
東銀栗田ビル　外空間　1978 年
NTT 広島仁保ビル　外空間　1980 年
工学院大学八王子校舎 5 号館〜 11 号館群　外空間　1986 年
工学院大学八王子校舎 3 号館　外空間　1987 年
スコープ本社 GLASS HOUSE　外空間　1990 年
以上の 10 空間である。

　これはどういうことなのであろうか。使われている素材がそうさせるのか、あるいは構成と展開が原因なのであろうか。私にはそれぞれの空間が醸し出すやすらぎとやさしさが、時を経て、そのような移ろいを感じさせるのではないか、と考えてはみるけれど、このことは私にとって、ものを創り出す以上に難しい課題なのである。私は考えあぐねてこのことを次のように自身にいい聴かせて納得を計る。この 10 の空間は、私の内部にそれぞれの空間を創り出す度に、君臨するのであろう超越者の思召によるのであろう。即ち超越者はこの間、常に加護の手を降し下さっているのではなかろうか。

　その超越者は神と同じように、私の正面やや右手上方 30 度に現れると自覚している。

　こういう表現をすると、「私は狂気と正気の挟間で仕事をしている」と随分と気取ったことを言っているな、との誤解を受けるやも知れぬ。然し私はこれは仕方ないことだと諦めている。私でさえ、時折、自分自身がさっぱり分からないのだから。

　このビルの外部空間は、図のような建物の東・西・南・北の四面を囲む各々が方形の合計面積 800 ㎡程の空間である。このような形の空間は、それぞれを図のような高低差を持った直線構成として細分化し、展開の中で一つの語りを謳い上げることがよかろうと構成と展開の方針を決めた。

　敷地西側では、エントランス部分、地下駐車場出入部

分などを除くほかは、公道（歩道）がふくらんだ形で、低い（H1300mm）石積の上に落葉樹林を展開、ビル周辺に介在する人びとの心を快適に抱き込むような展開とし、他方、この西側公道からビル東側外空間までの眺めを連続させることを考え、エントランス階段の天端を公道境界から1270mmの高さとし、公道を、ビルファサード前を往還する人びとにも快適な眺めを提供出来るよう計った。この1階ホールに介在する人びとと西側の外部に介在する人びとに快適な眺めを供している東側外空間のカスケードと滝の水の流れと落水は、この空間の中心部で、視的緊張感を大いに盛り上げている。カスケードと滝の構成と展開は、1975年小西酒造本社の5階の貴賓室の内空間に創り、建築家の柳沢孝彦さんによって「回路」と名付けられた、私の作品からのヒントである。

三番町KSビルの外空間は1991年11月に完成した。完成した空間をじっと眺めていた私は、ふと20年余りも経たろうかギリシャ本土・クレタ島を旅した折のハギア・トリアーダ、あるいはフェイトスの遺跡での空間体験が、こういう空間を創らせたのではないかと、遠く去って往った感動の日々と、もの創りとして受けた数々の暗示の日々がなつかしく思い出された。

この外空間は、参考となるであろう図面と写真を豊富に載せてあるので、じっくりと見て戴ければ幸である。それだけ得る処が多いと思うが。

完成1年後の1992年11月、北隣地の宮内庁用地と三番町KSビル敷地の境界辺りに弥生人の住居跡の発見を聴く。九段坂付近坂上は弥生人のムラが多いはずと聴いていたから驚くこともなかったが、最初にこの現場を訪れた日に漂っていた霊気は納得出来た。

やがてこの空間も人間の生活と深く関わり合った遺跡の匂いを感じさせるのであろうか、私は迫力ある10番目の外空間をこの三番町に生み出し得たと喜ぶのは何時の日であろうか。

私が私なりの空間を創り出す折に、私の内に問うことは、"もの"が空間でそれ自体独立して存在することはまずあり得ない。ものは他のものと関わり合いを持ち、規定され、条件付けられる。そうでなければ、そのものが存在する充足理由が成り立たない。そこに空間関係の位置が生まれ、時間関係の継起が成立する。けれども、この2つを論理的に把握することはまず不可能である。それを把握する唯一の方法は、純粋直観だけであろう。このように空間における"もの"の存在ということの難しさを考えると、資質という問題が大きな要素となる。

この辺りで私は超越者を登場させる。この超越者とは、私にとって神なのか何なのか。

純粋直観は、私にとって超越者の私への加護の手であると表現する以外に解決の方法がない。

1991年
用石の種類
石積（笠石共）　　　　黒みかげ石ラステンバーグ
カスケード縁石、床　　〃
流れ床　　　　　　　　〃
転石　　　　　　　　　〃　　　、芦野石、中国産みかげ石
化粧土留ボーダー石　　〃　　　、芦野石
舗床板石、ポーチ板石　稲田石、中国産ピンコロ
床鼻石　　　　　　　　〃
窓石　　　　　　　　　〃
階段　　　　　　　　　〃
擁壁、17柱　　　　　コンクリート打放直仕上（ツツキ）

※詳細図はディテール図面編の182～187P参照

Sanban-cho KS Bldg. '*Gai-Kukan*' 1991

It is one of the most beautiful scenes in Tokyo today that reflecting moat of water surface which Surrounding with Koukyo-Gaien (the moat of the Imperial Palace), the great stone ,reataining walls, and on the place where through for Soto-Sakurada-Mon, Miyake-Zaka, Hanzomon, Chidorigafuchi, its reataining walls, old pine trees, with the modern buildings of the surrounding areas .
Very near the place, the KS building standing.
The Sanbancho, standing KS building is placed through Yasukuni-Dori from Kudanzakashita to Shinjuku turn left Embassy of India on Kudanshita , and through Uchibori-Dori against Miyake-Zaka near the Nishogakusha-University on the way.
 It stands in between Chidorigafuchi, known as cherry-blossom viewing and Uchibori streets. It is Surrounding Fairmont Hotel, Imperial Household Agency sites, Cemetery for war dead, Soga-residence adjacent KS building, across the street Tobishima Corporation. In a part of Tokyo that is well known for the amount of greenery it has. One of the neighboring plots is the Chidorigafuchi graveyard, which has a lot of trees and adds to the quiet atmosphere of the area. The KS building is ten stories tall and is covered in granite from Korea. The design of the building is novel and attractive and due to the way it was sited. I conceived here is fitting element of my concept of which 'creating a spaces that are unified with a road'. Namely, seemingly the road swells, forming a space that embraces the people there comfortable, using the surrounding building facades as an effective backdrop fundamentally, refer before text.
I was delighted with the prospect that I could produce perhaps the most comfortable space ever.
Furthermore Strangely I felt the Spirit or Aura atmosphere of which I felt like Kudanzaka-shita when I visited for designing '*Gai-Kukan*' of the Hotel Grand Palace two decade before. Surprisingly Whenever I felt this kind of atmosphere before, design well and eventually could produce a powerful space. It is impressed the people visiting here and nearby, giving pleasant feeling of encounter with historic remains. Powerful space means, I believe, a space produced by people and that unites these objects with people in a meaningful way.

What is the reason for this? Is it for causing materials or the composition and development of the design? I wonder if it is because the restful and tender atmosphere of the space makes people feel such a lapse of time, but this question is more difficult for me to find an answer. It is easier for me just to create things.

 I am trying to satisfy myself by which the transcendental existence inside me it is governed the paces and it presides over me every time. The transcendental existence must have given me a hand for help in producing these spaces.
I perceive this transcendental being as it appears, like god, in front of me about 30 degree to the right and upward.

When I say things like this people say I'm working on the border of sanity and insanity. I'm really not too sure myself sometimes. As shown in the figure, the building has rectangular spaces on all sides, amounting to about 800 sq. m. in all. My basic principle for the design and composition was to devide these spaces with straight lines into separate parts of different heights. In order to allow a free view from the street on the west side to the space with the waterfall on the east the entry steps were limited to a height of 1270mm. This plays a central role in linking the spaces and offering visual stimulus to the pedestrian level. The '*Gai-Kukan*' was completed in 1991. It seems my experience at the Hagia Triada and Feist on a trip to Greece and Crete two decade before played an inspirational role in the design of this space . Also, one year after the completion of this project, I learned that some remains from the Yayoi period were found in this area. Perhaps that explains the spiritual sensation I felt on coming to this site for the first time. When I create a space I assume that no one object can exist independently. Only through interactions with other objects, through mutual restraints and restrictions, can an object have true meaning. Thus spatial relations and connections are born among objects. It is impossible to explain such interactions logically only intuition can allow one to understand such a situation One's natural disposition becomes the main point here. It is here that I believe the hand of a transcendental existence can be seen in my work.

1991
List of stones
Stone wall (including capping stone): Black granite 'Rustenburg'.
Edge of the cascade, paving: same as above.
Stream paving: same as above.
River cobble: same as above. Ashino-shi. Chinese granite.
Ornamental edging stone: same as above. Ashino-ishi.
Paving flag stone, paving for porch: Inada-ishi. Chinese granite setts.
Tokobana-ishi: same as above.
Mado-ishi: same as above
Stairs: same as above
Retaining wall, 17 pillars: Cast concrete with finished with prodding.

※ details 182-187p

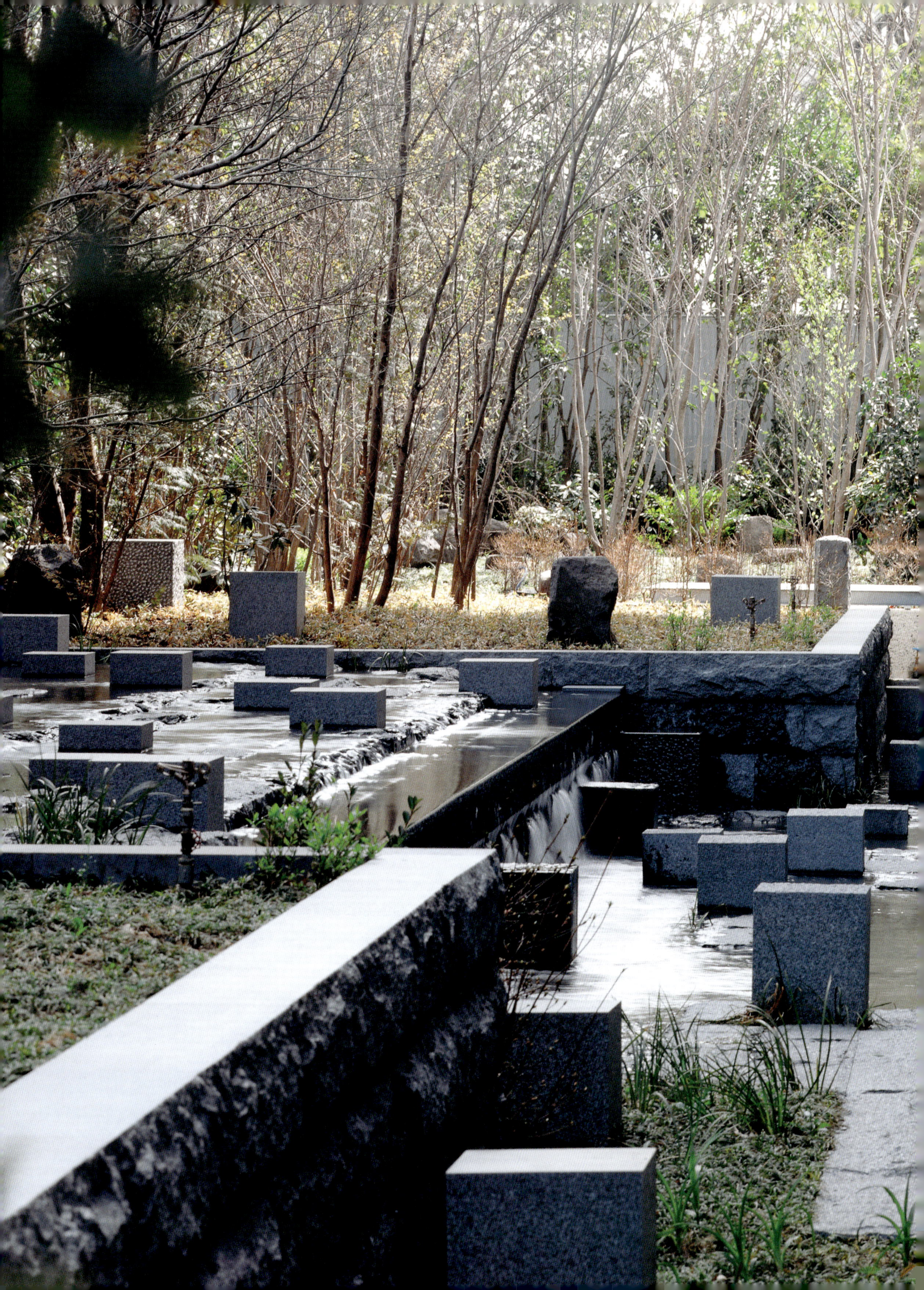

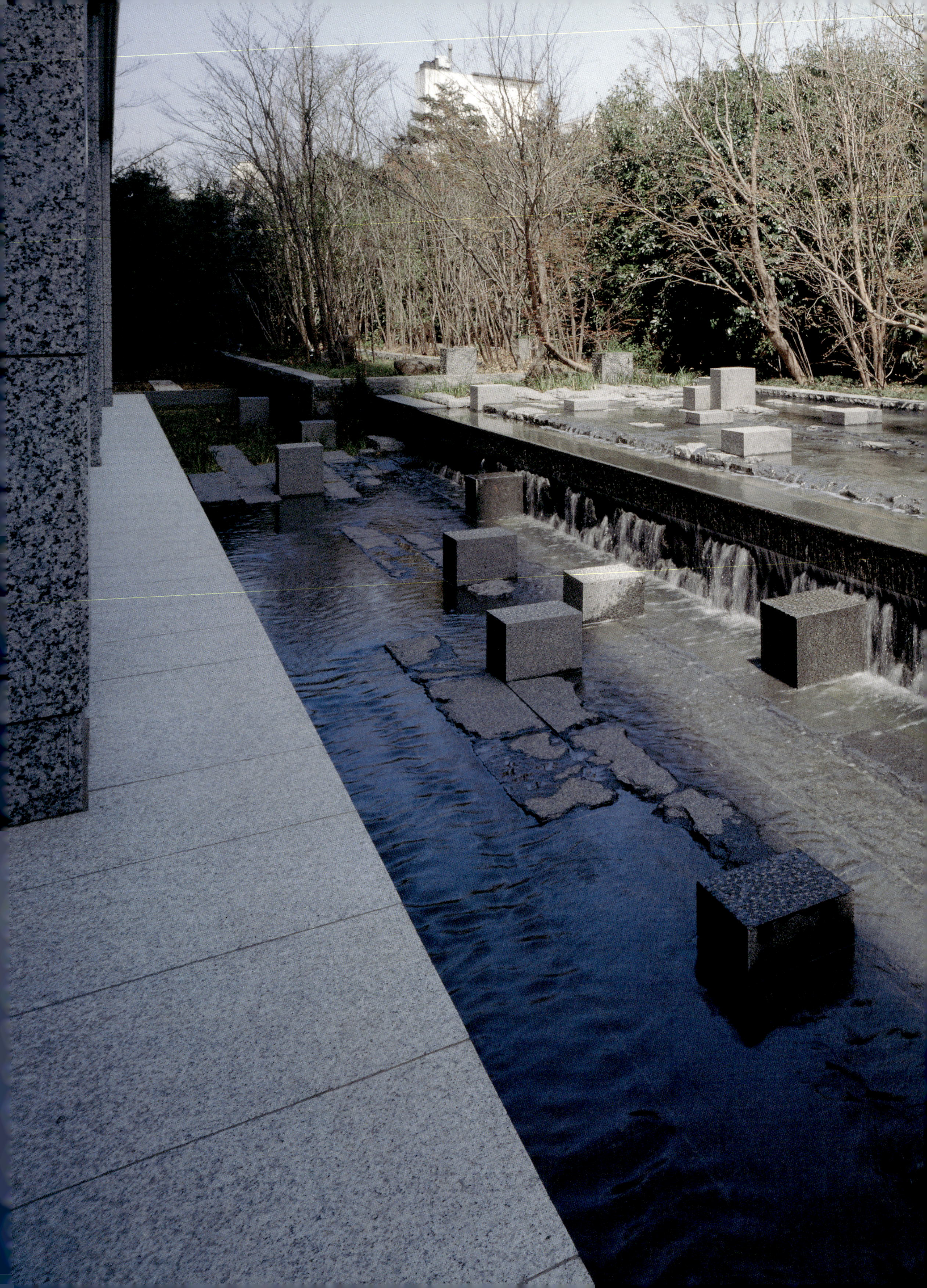

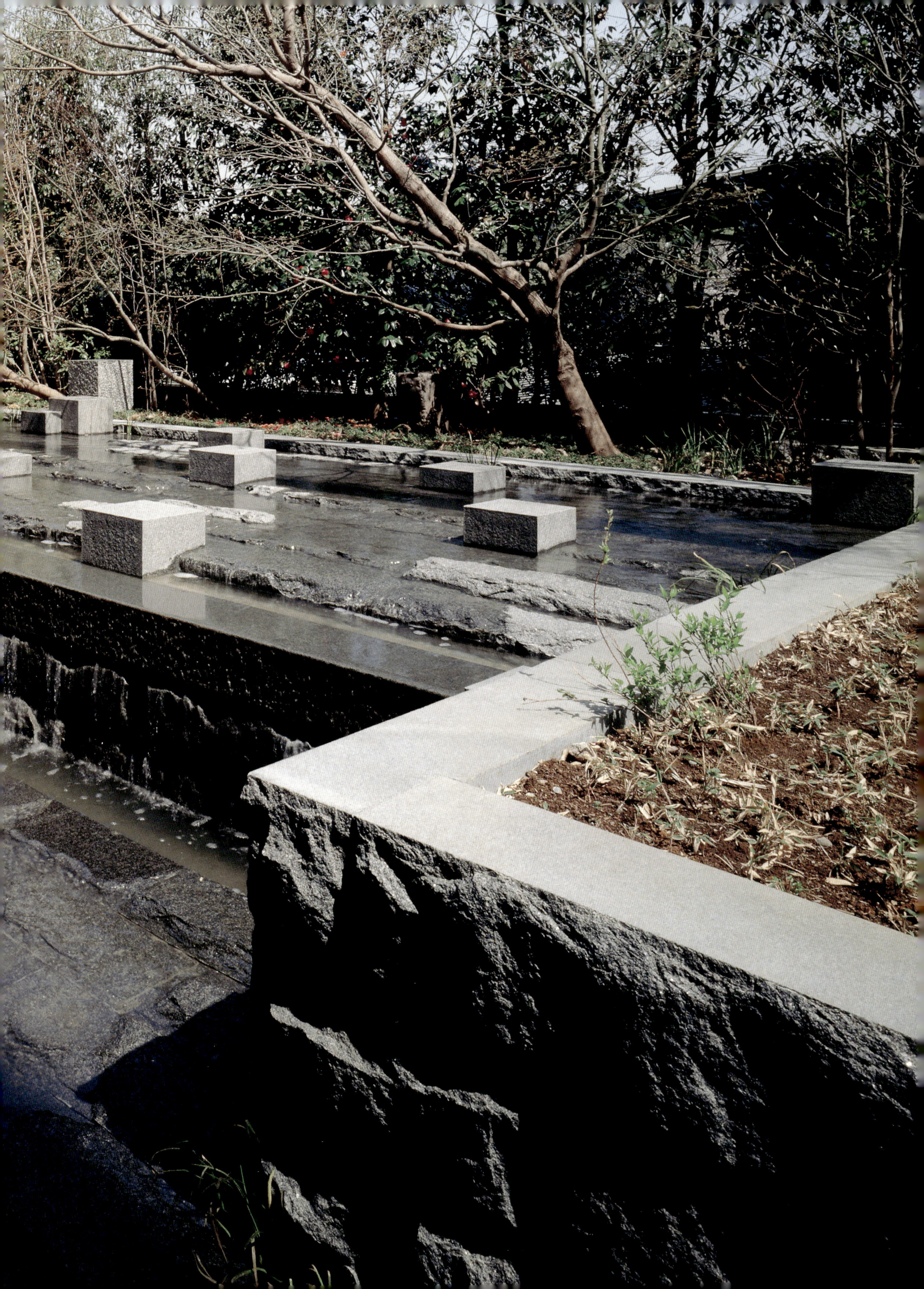

146

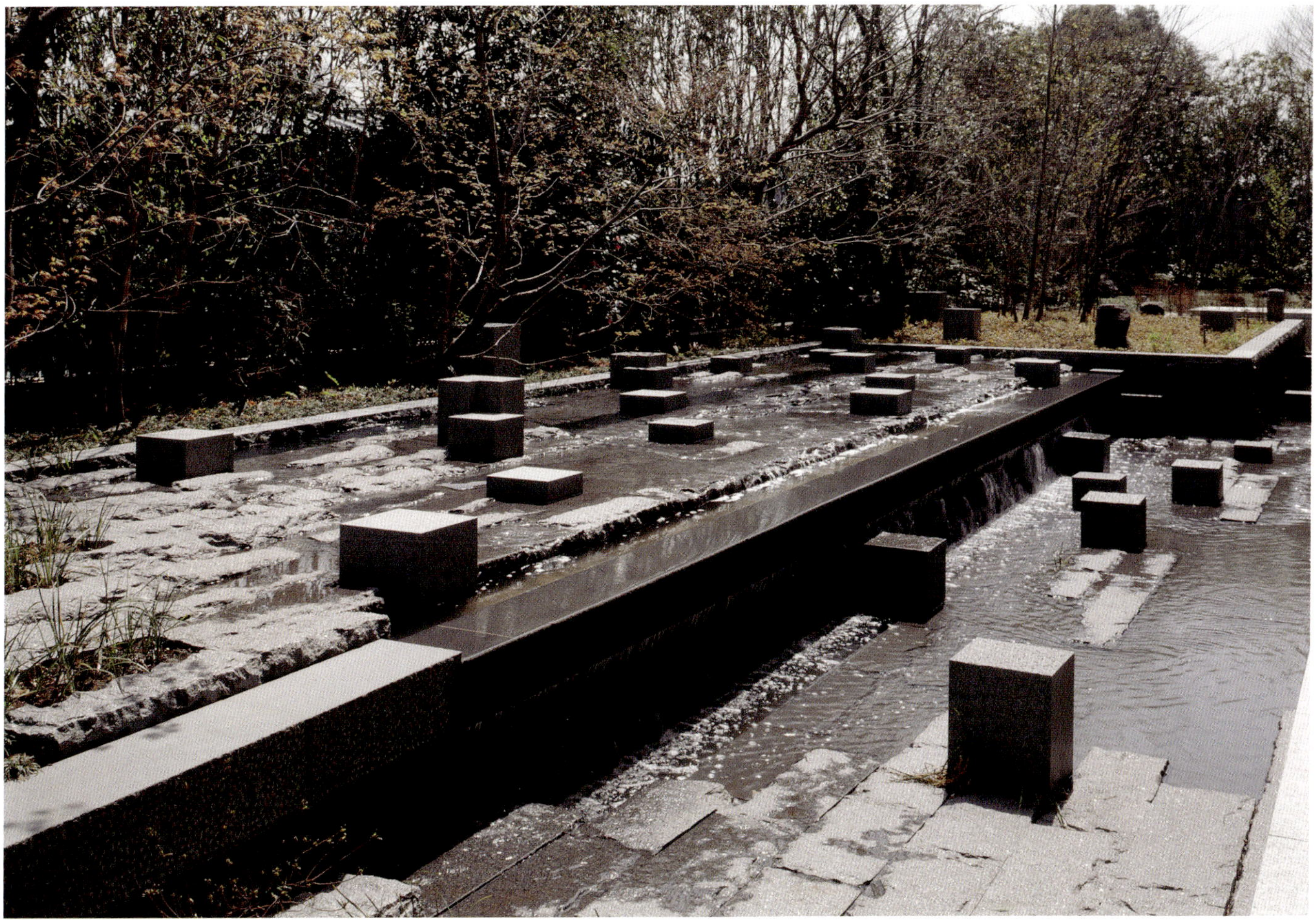

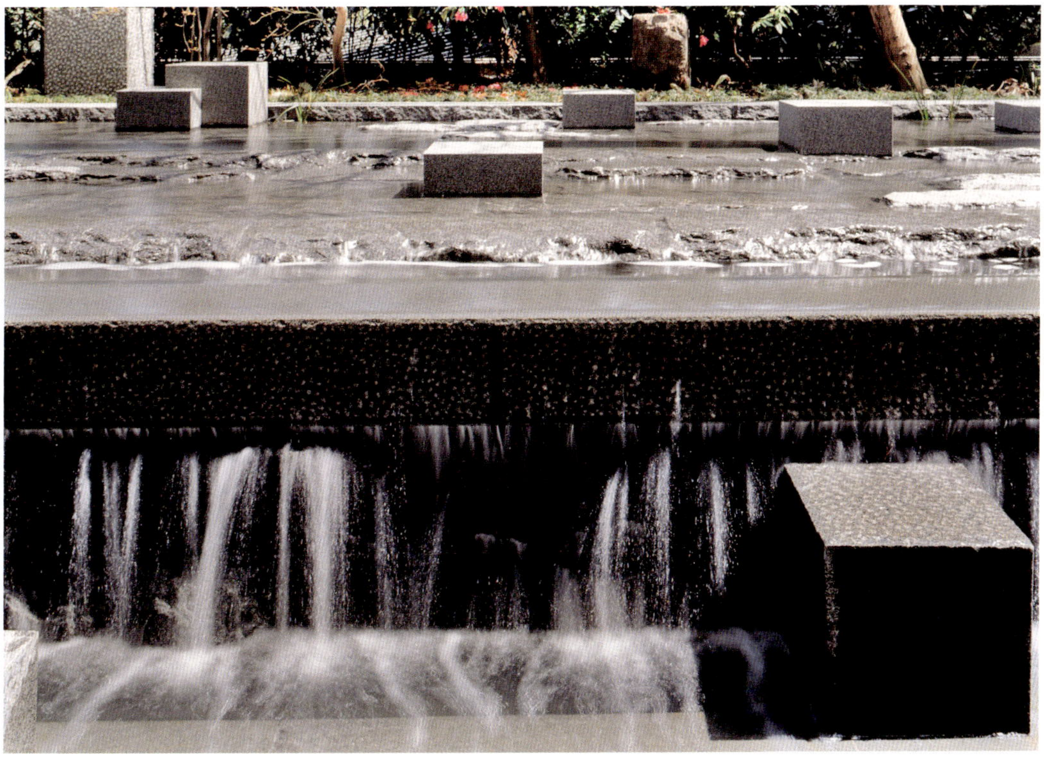

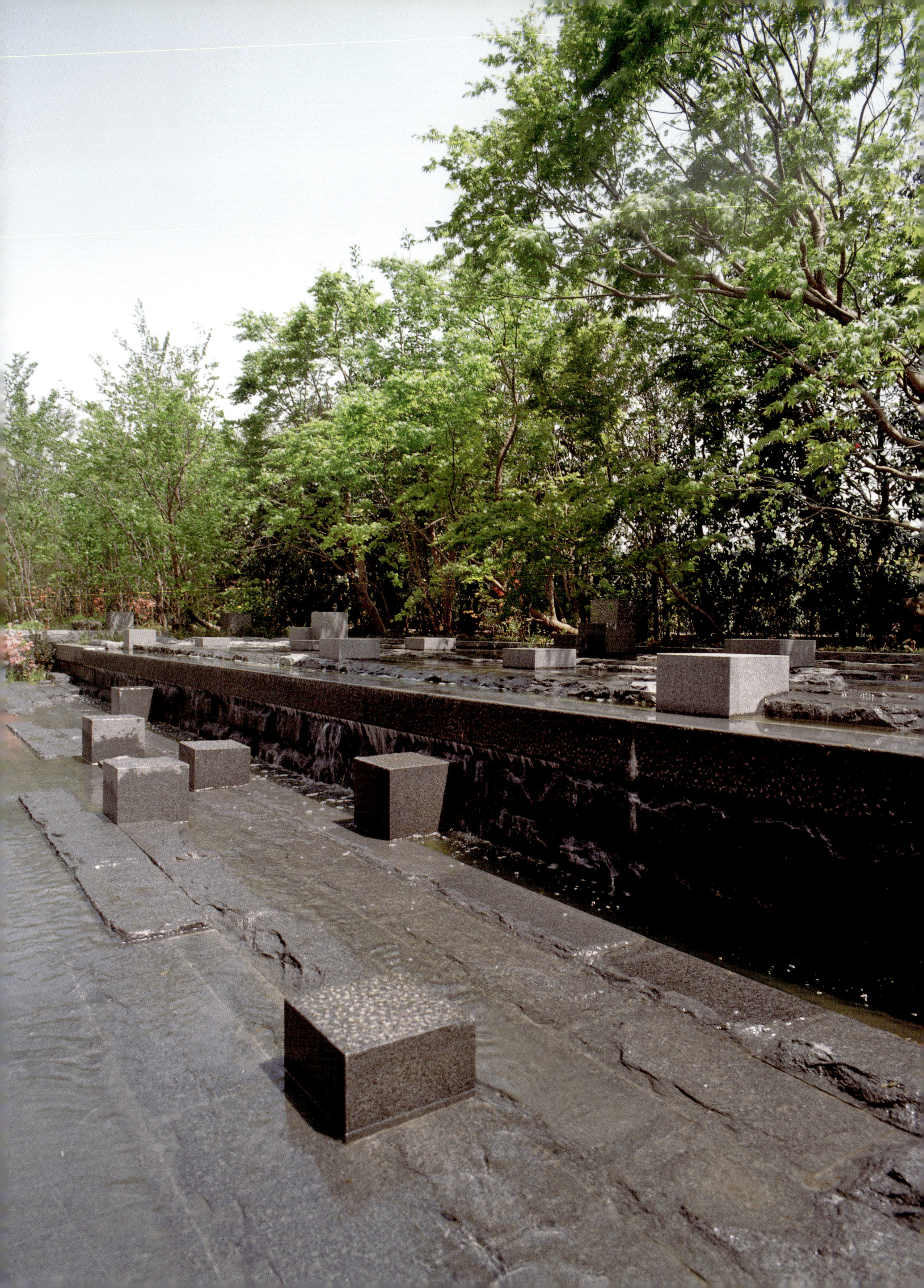

149

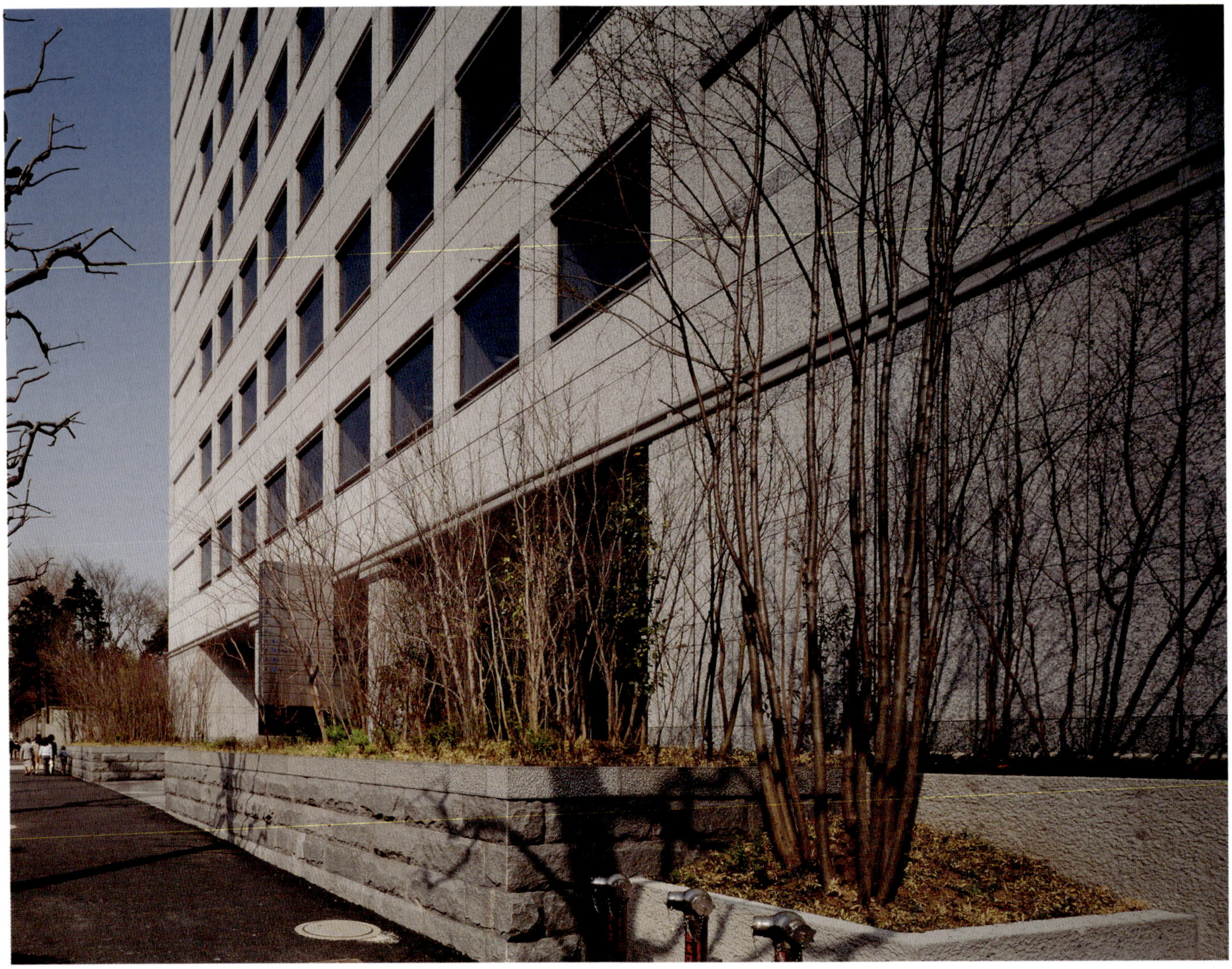

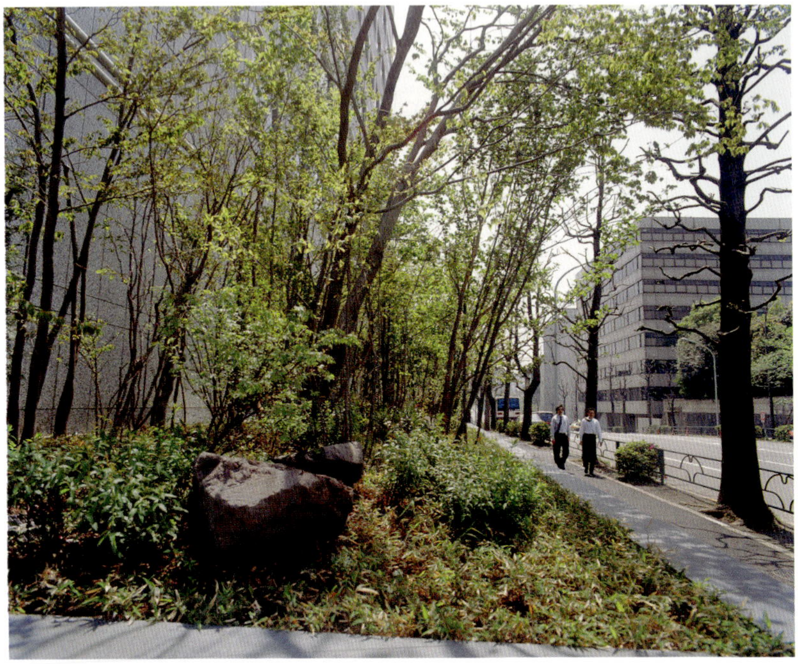

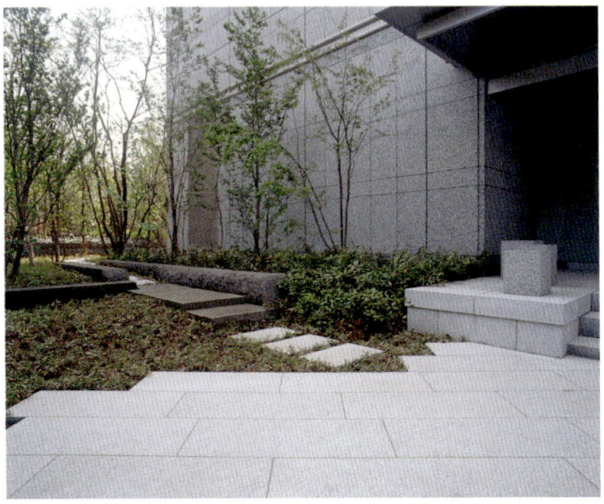

第2章　Chapter.2
創作ノート　Notes on Creation

立っている精神 1979
Spirit of Standing Still 1979

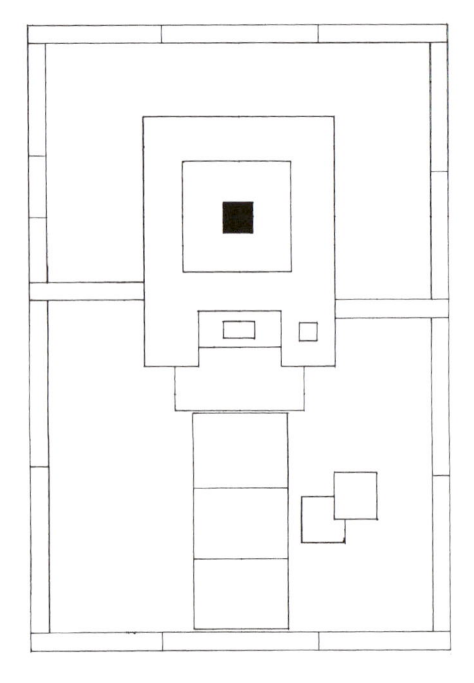

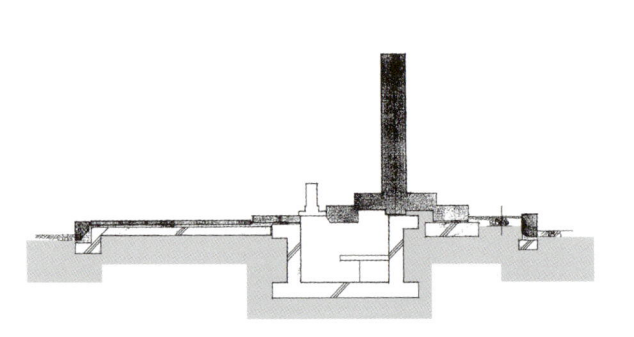

平面図 1/60

断面図 1/80

152

哲学者の矢内原伊作氏には数多くの著作があるが、1970年代に京都の雄渾社から"矢内原伊作エッセイ"として、『海について』『現代人生論ノート』『文学論集』『芸術家との対話』ほかが出版されている。その70年代の後半になろうか、同じ京都の淡交新社から『石との対話』と名付けられた写真入りのエッセイ集が1冊の本として出版された。この『石との対話』は、もの創りの私にとって"心で見る世界""心で創る世界"など、心（精神）とは何かを失いつつあった難しい時の流れの中で、多くの示唆を与え、導かれるものが多かった。

その『石との対話』の中の「立っている精神」の章に次のような記述がされている

「地面に横たわっている細長い石はなんの感興もよばないが、垂直に立っている石はわれわれをおどろかす。眞直に立っているだけで、それはもう精神のしるしである。それは標柱であり、記念碑であり、あるいは墓である。立っている柱は、それは均衡だが、いわゆる危険にさらされている均衡であり、不安定にうちかっている安定だ。柱の美しさは、ささえる力の美しさにほかならない。立っている不動だが、動に打克っている不動だ。逆に云えば、立っている姿勢はいつでも動くことの出来る姿勢である。横になっている人間は物体に近いが、立っている人間には精神の集中がある。」

上の記述は、永い間私の脳裡から離れることがなかった。1979年、人伝に、日本銀行出身で北海道銀行の創始者である故島村融氏の墓標その他を京都の宇治平等院の墓域に建立することを依頼された。無神論者・故島村融氏の墓標は、「立っている精神」の記述に彷彿されるような姿として、写真と図面に表した。

「立っている精神」はその後、私の内に消化されて、1986年工学院大学八王子校舎5号館〜11号館群の小広場に3本の石の柱としてその心（精神）を表している。

Among many other works written by philosopher Isaku Yanaihara, 'About the sea', 'Notes on Modern Life', 'Literary Essays', and 'Dialogue with an Artist' were published as 'Essays from Isaku Yanaihara' by Yukonsha publishers in Kyoto in the 1970's. Later, probably in the late 70's, another book of essays with photographs called 'Dialogue with Stones' was published by Tankoshinsha publishers, also in Kyoto. For my creative self, this book, the 'Dialogue with Stones' inspired me enormously with suggestions like "the world seen through the mind" and "the world created in the mind", at a time when things were spiritually confusing.

The following is a quote from the chapter "The Spirit of Standing Still" in the "Dialogue with Stones".

A long and narrow rock laid on the ground would not be impressed us, however on the contrary, an upright rock is striking. It is, so to speak, a symbol of the spirit itself, just by being perpendicular. It is a milestone, a monument, and yet also a gravestone. The stone pillar itself is balanced, however while it is a balance at risk its stability overcomes instabilities. The pillars' beauty is that of support. The stillness prevails against movement, although the stillness is standing upright. Conversely, a standing posture is ready to move on at any time. A body lying down is similar to a still object, but a standing body possesses a spirit of concentration.

The description above stayed in my mind for a long time. In 1979, I was commissioned to construct a gravestone in the Uji-Byodoin graveyard in Kyoto for the late Toru Shimamura, who had formerly been with the Bank of Japan and later was the founder of The Hokkaido Bank. As the late Toru Shimamura had been an atheist, his gravestone was built to suggest the description of the "Spirit of Standing Still", and it is evident in photos and drawings.

Afterwards, the "Spirit of Standing Still" was digested in my mind, and the spirit is embodied in the three stone pillars erected in the open area among campus buildings No.5 - No. 11 of the Kogakuin University, Hachioji, in Tokyo in 1986.

厚別商業センター 外空間 1977
Atsubetsu Shogyo Centre *'Gai-Kukan'* 1977

京王プラザホテルの外空間から6年ぶりで都市の公共空間に割り込んだ空間が生まれた。

　昭和51年秋から52年春にかけての新札幌駅前の厚別商業センター外空間である。

　この空間は本来三菱地所の設計によることになっていたが、途中から私が変わって担当した。非常に厳しい工費であったが、センタービルの外装からペデストリアンデッキの仕上げに到るまで変更して費用を捻出し、綜合的にまとまった出来上がりとなったのは幸いであった。

　舗装のタイルはペデストリアンデッキを含めて、伊奈製陶の売れ残り在庫タイルを格安で分けて貰い、これとコンクリート打放の壁（直仕上げ）で大部を構成ポイントに子供の遊び用として、小さな泉などを設け、空間からの働きかけをつくった。厳しい予算と面積の広いことから考えてみるとまあまあの出来と考えている。

　けれども二度とタイルの床の広場とか舗装はつくりたくはない。

　広場の左側に北海道銀行の支店がある。2年後の54年に私はこの銀行の創始者 島村 融氏の墓を京都府宇治市の平等院の墓地に建立した。これも何かの浅からぬ因縁であろう。

Six years after the 'Gai-kukan' for the Keio Plaza Hotel, a space squeezed into an urban public space was born. It's a new 'Gai-Kukan' for the Atsubetsu Shogyo Centre in front of Shin-Sapporo Station due to open between fall 1976 and spring 1977.

This space was originally going to be designed by Mitsubishi Estate Co., Ltd., but I was put in charge of it while it was in progress. It was lucky that I managed a unified look by modifying the existing plans from the exterior of the Center Building to the pedestrian decks, and devised to cope with the costs on a very limited budget.

All the paving tiles including those for the pedestrian decks were bargain priced, unsold products from Inax Co., and also the cast concrete wall was used widely as a focal point and play area for children complete with a small fountain, and I aimed for some action from the space. Considering the size of the large site and tight budget, I think I did pretty well.

Having said that, I never want to design a large public square of paving tiles or stones again.

A branch of The Hokkaido Bank is situated on the left side of the public square. Two year later in 1979, I was called upon to design the gravestone of its founder, the late Toru Shimamura to be erected in Byodo-Temple in Uji-city, Kyoto-Pref. There must be some sort of close connection.

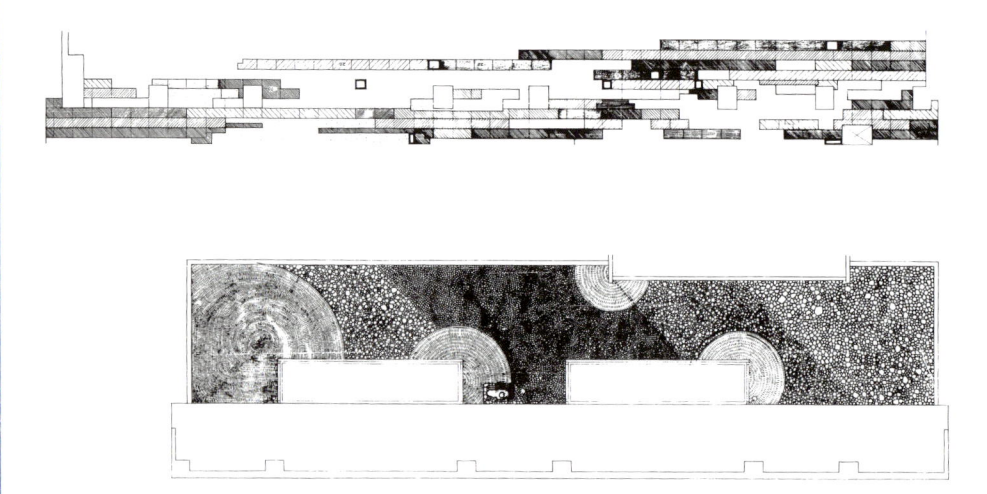

朝日生命成人病研究所 1968
The Institute for Adult Diseases, Asahi Life Foundation 1968

イトーヨーカドーグループ多摩研修センター 外空間 1986
Ito-Yokado Group Training Center in Tama '*Gai-Kukan*' 1986

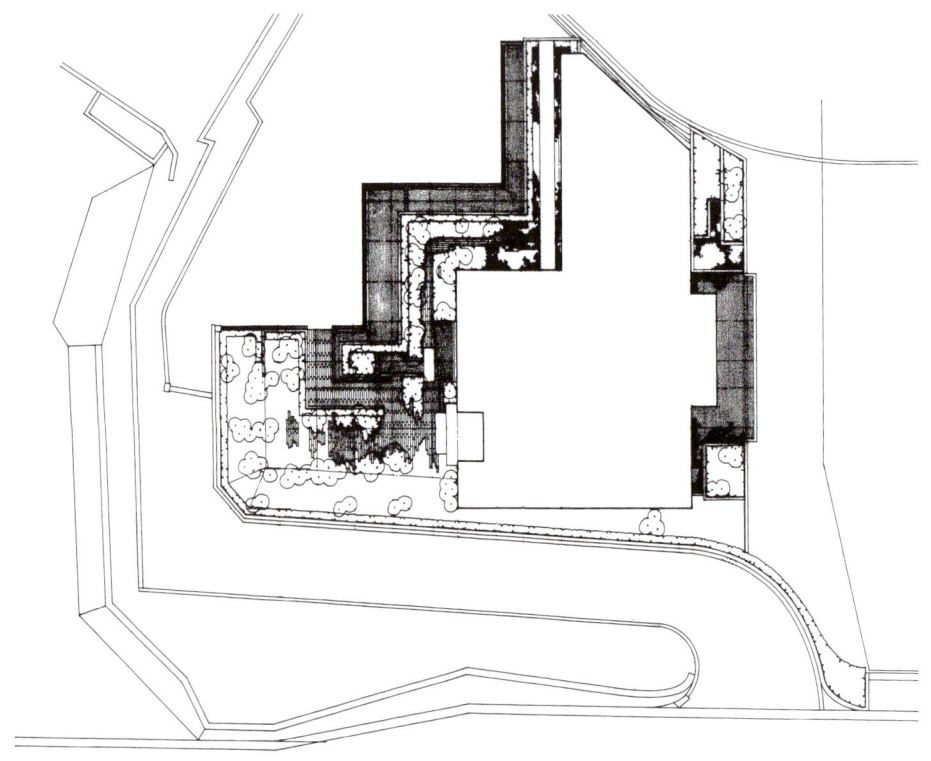

レストランあかべこ 外空間 1974
Akabeko Restaurant *'Gai-Kukan'* 1974

平面図 1/600

158

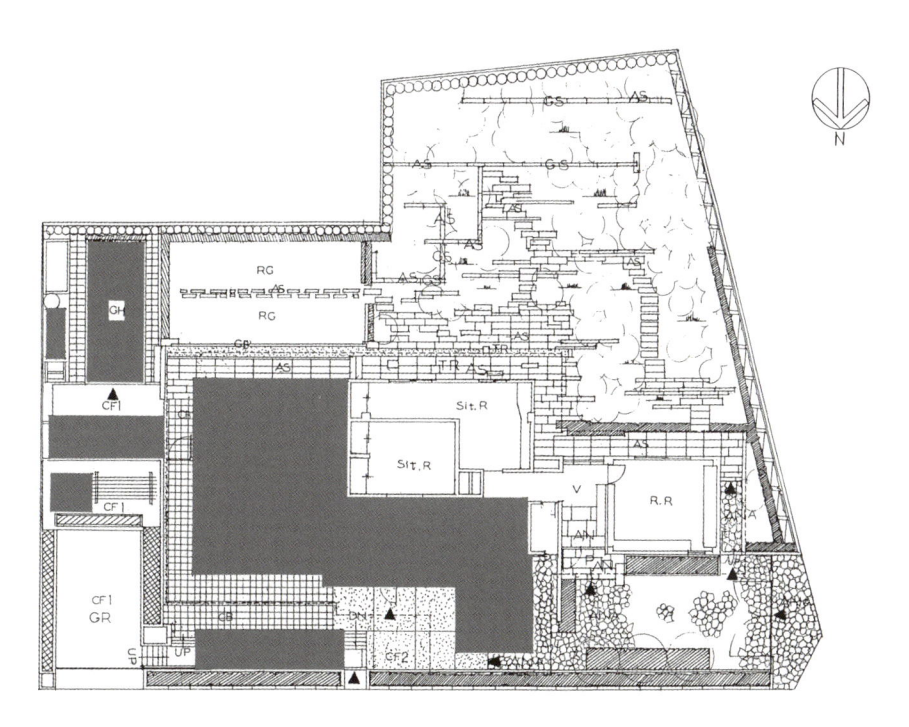

松方邸 1972
Matsukata Garden 1972

平面図 1/350

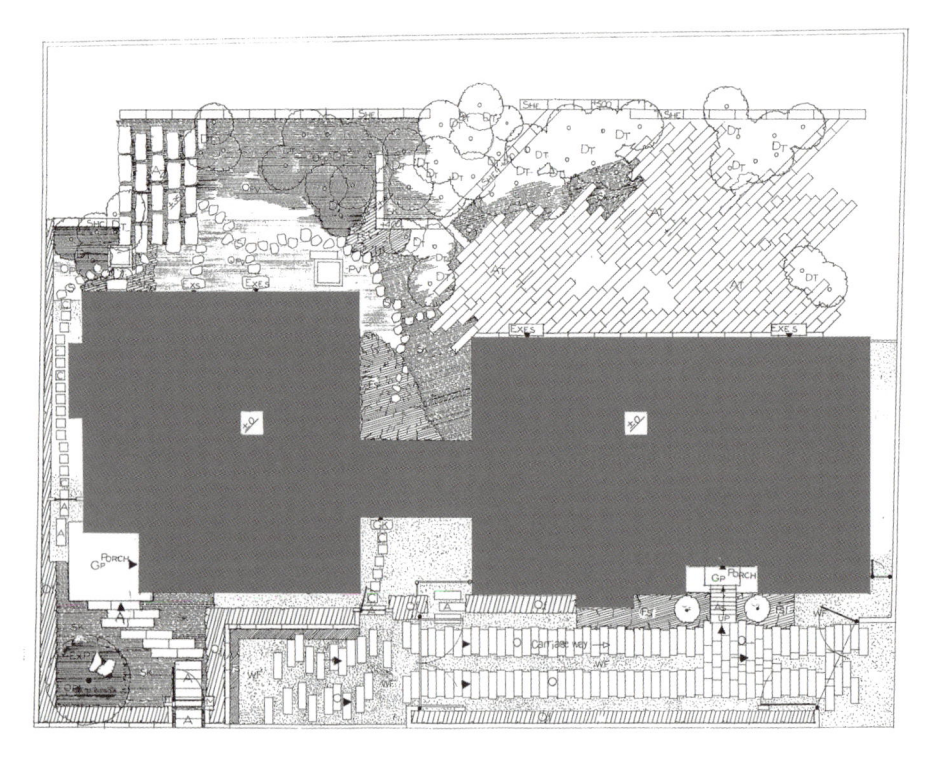

小寺邸 1976
Kodera Garden 1976

平面図 1/250

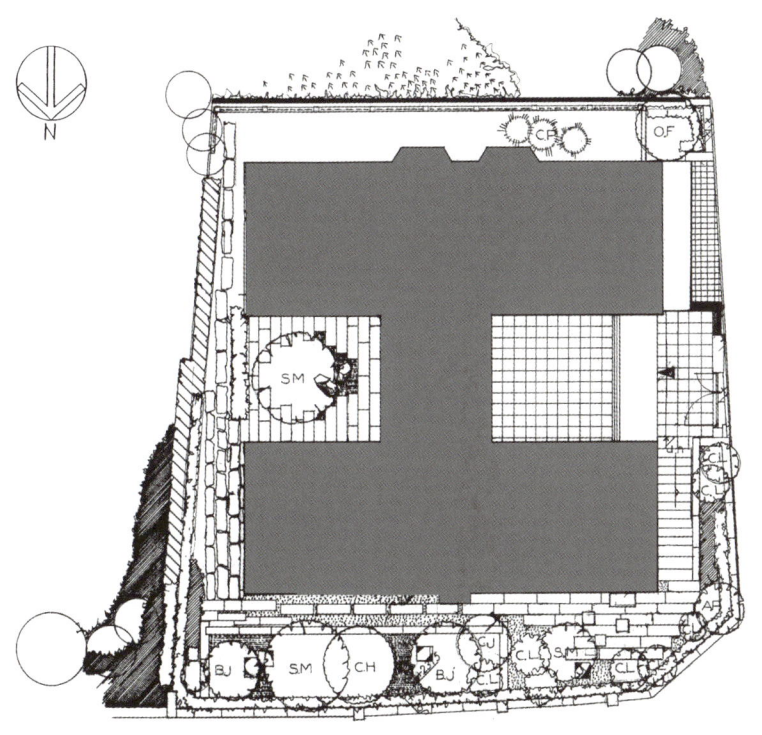

水野邸 1987
Mizuno Garden 1987

平面図 1/200

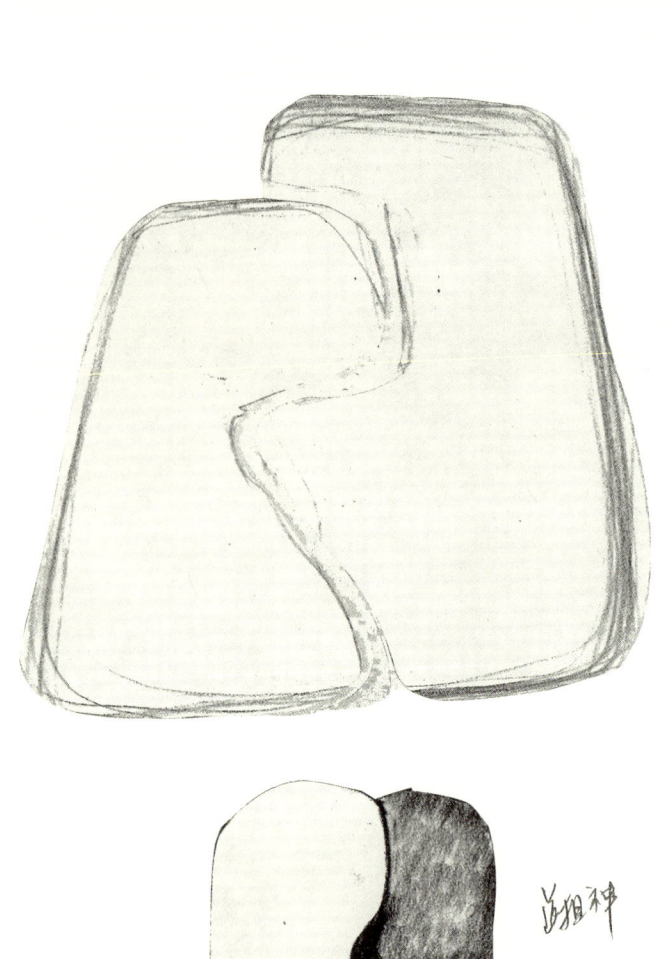

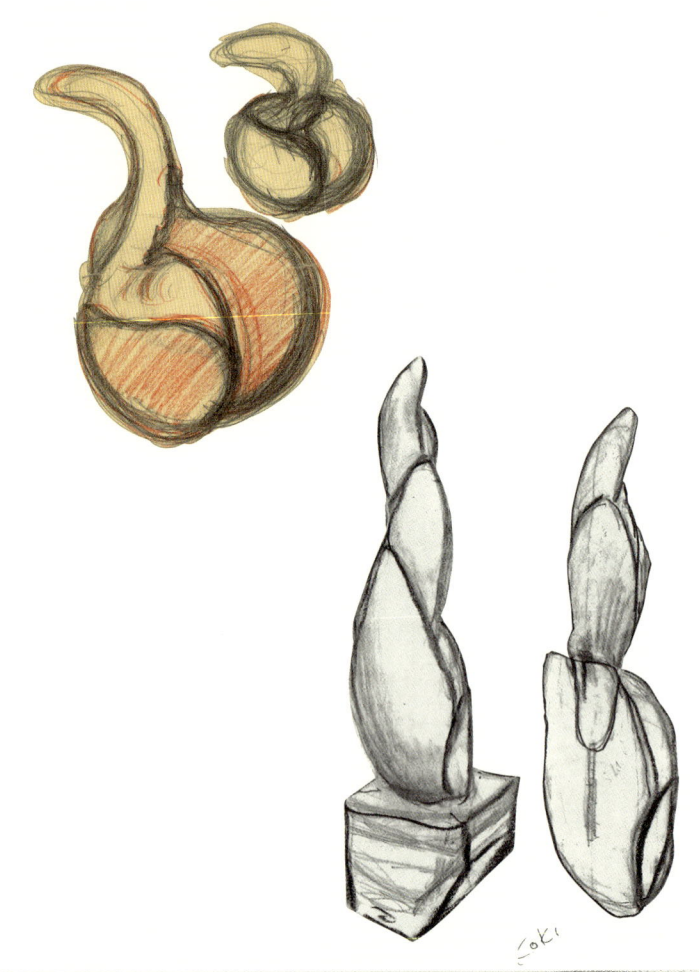

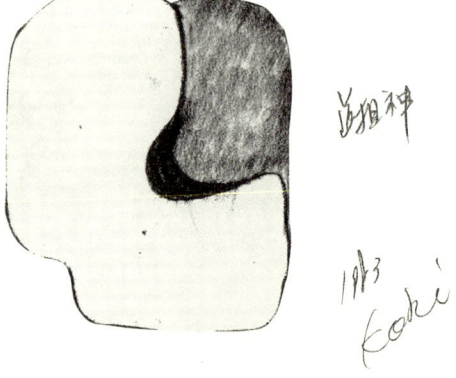

道祖神
1983
Koki

Koki

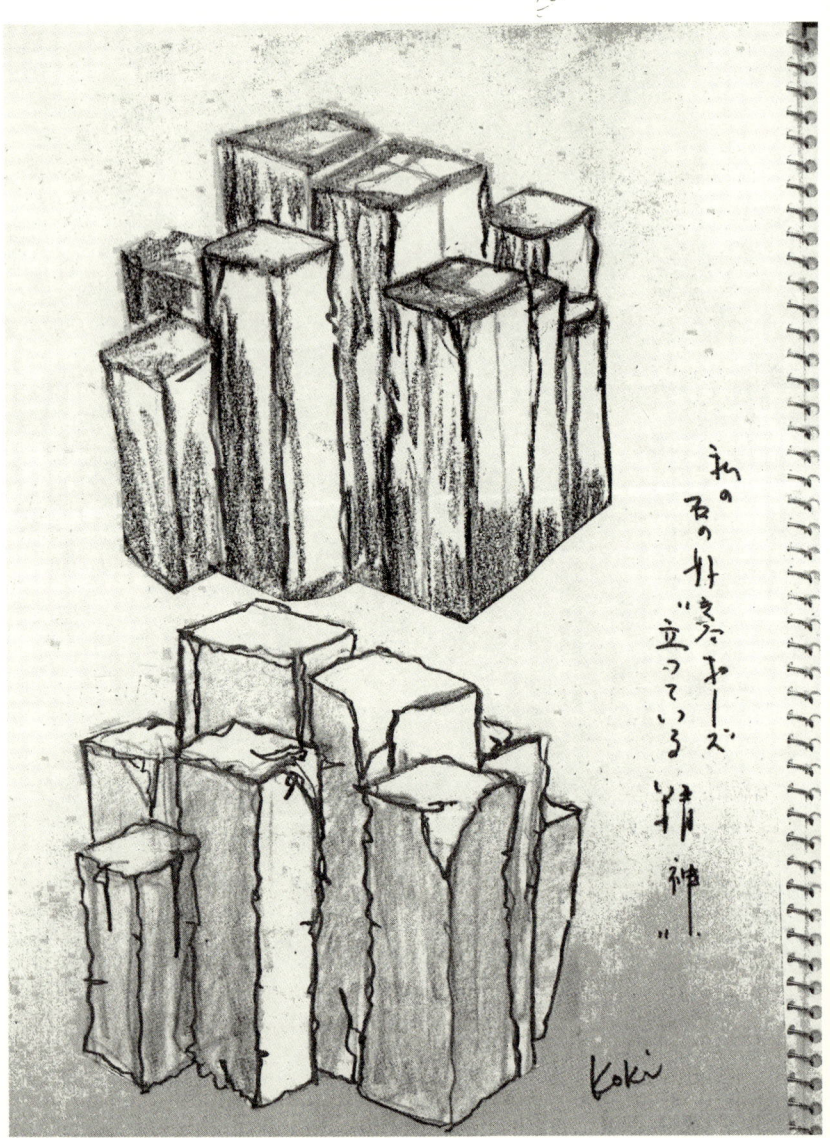

松林の精神
Koki

私の石の井桁ポーズ
立っている精神
Koki

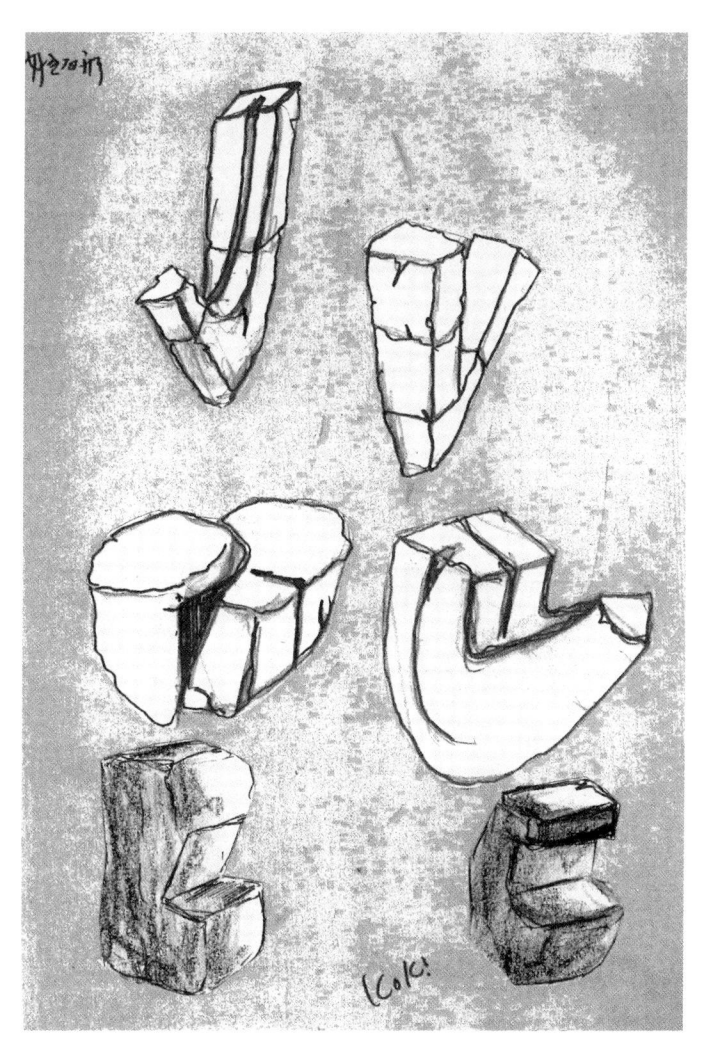

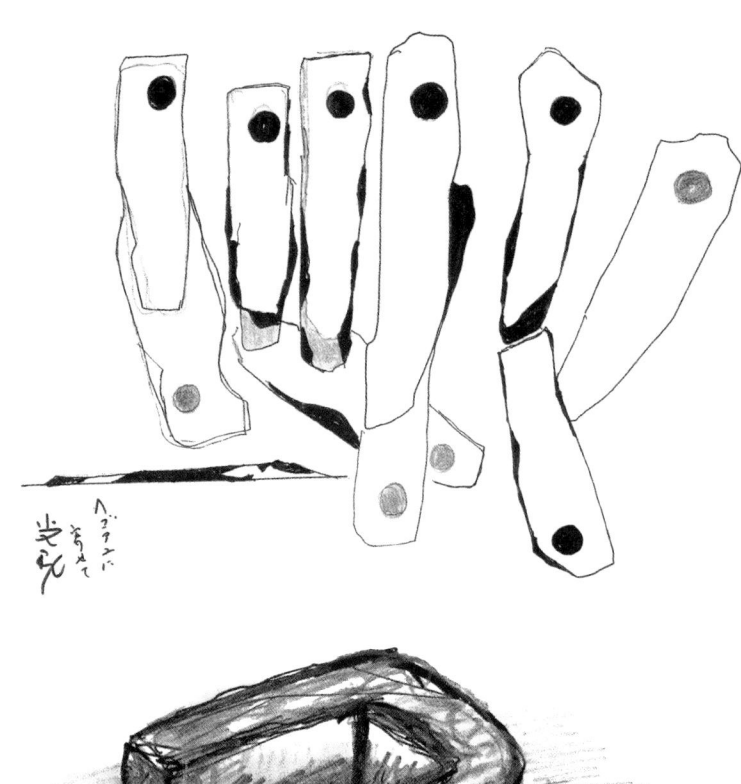

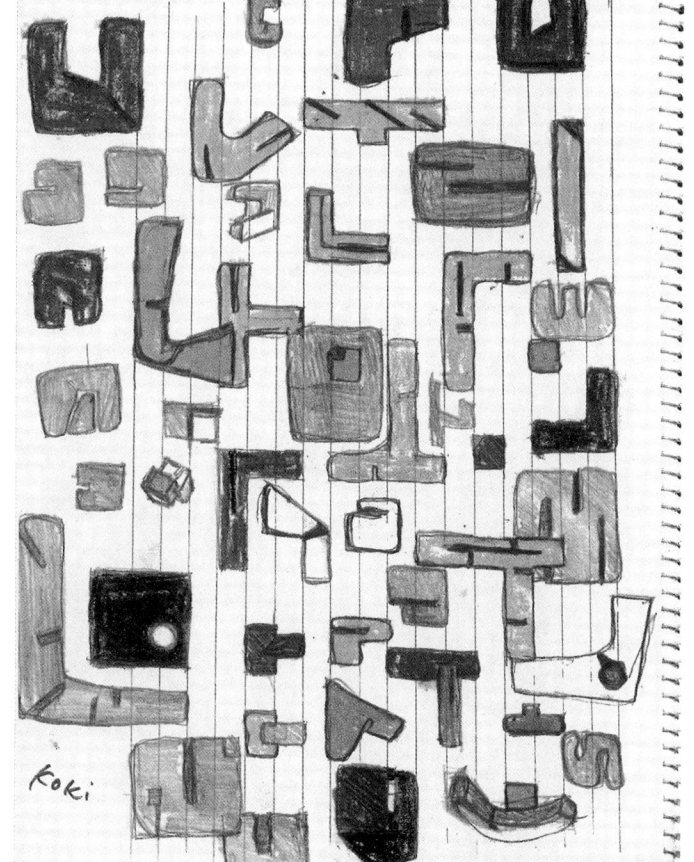

ある人に
"今何が作りたい?"と問われたら

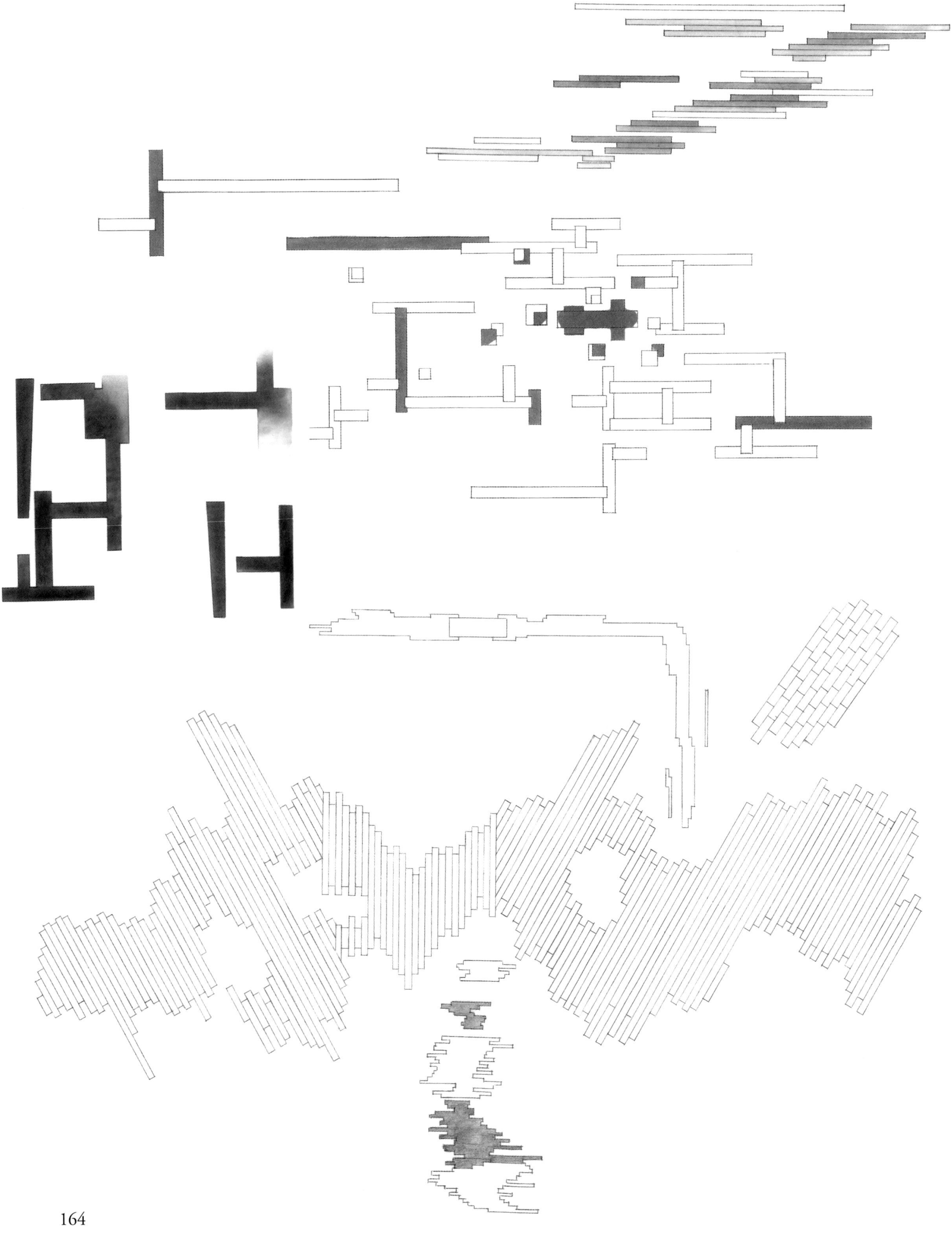

付録 Appendix

Details

ディテール図面編

京王プラザホテル
北館 4 号街路

平面図

C-C 断面図

E-E 断面図

C-C

B-B 階段詳細図

7階屋上庭園

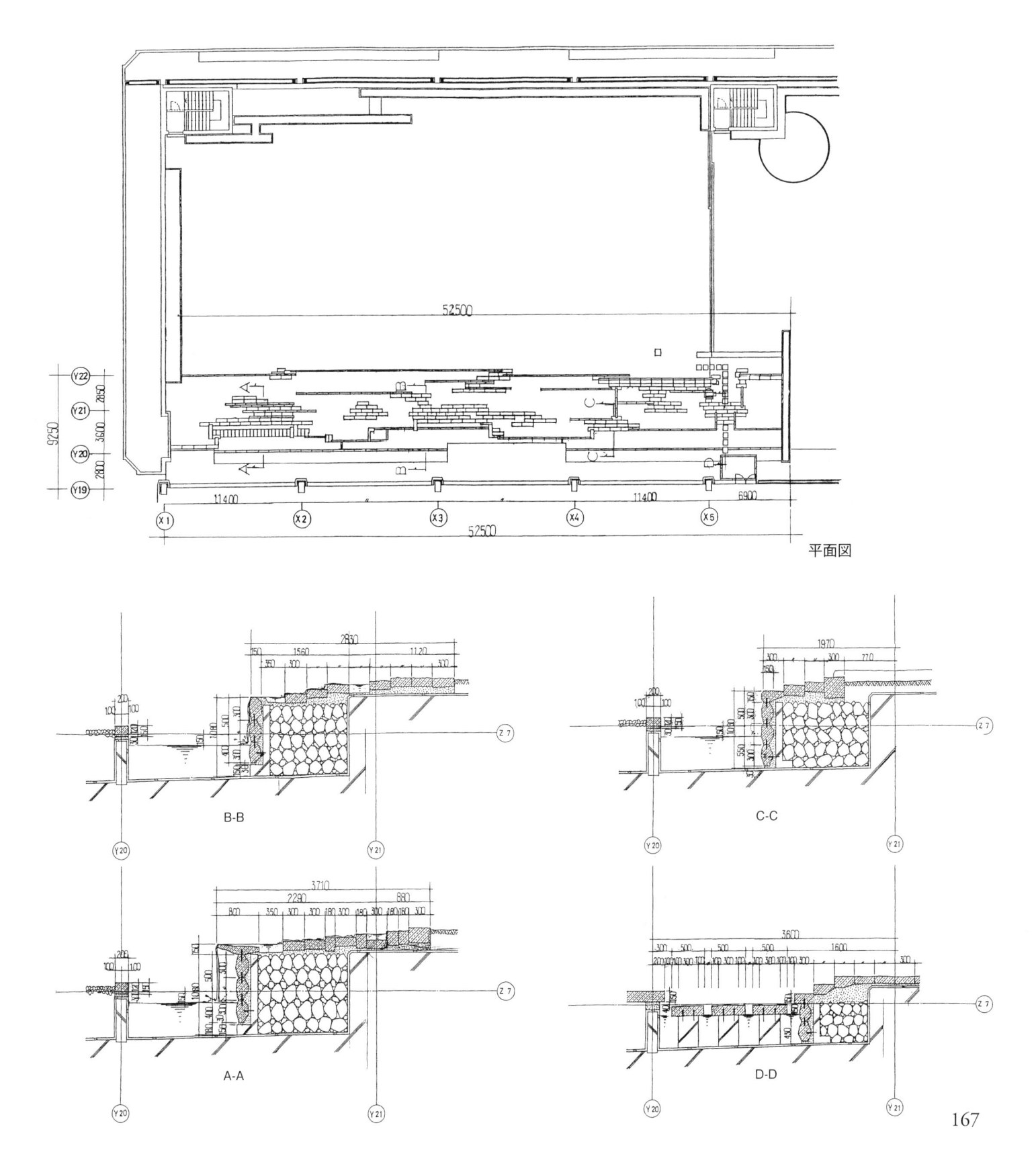

平面図

B-B

C-C

A-A

D-D

日野自動車工業本社

平面図
A-A
B-B
C-C
D-D E-E

断面図

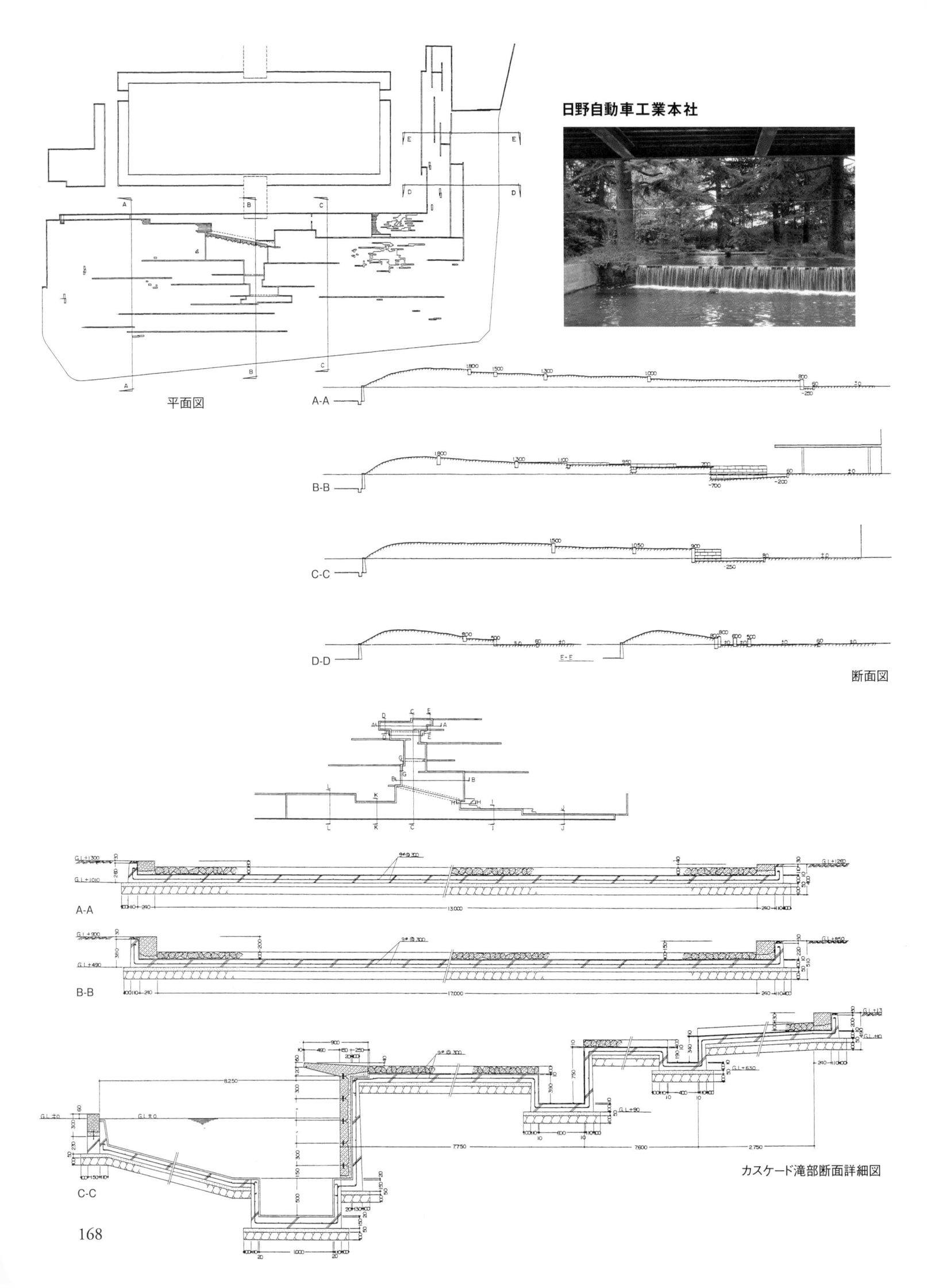

A-A
B-B
C-C

カスケード滝部断面詳細図

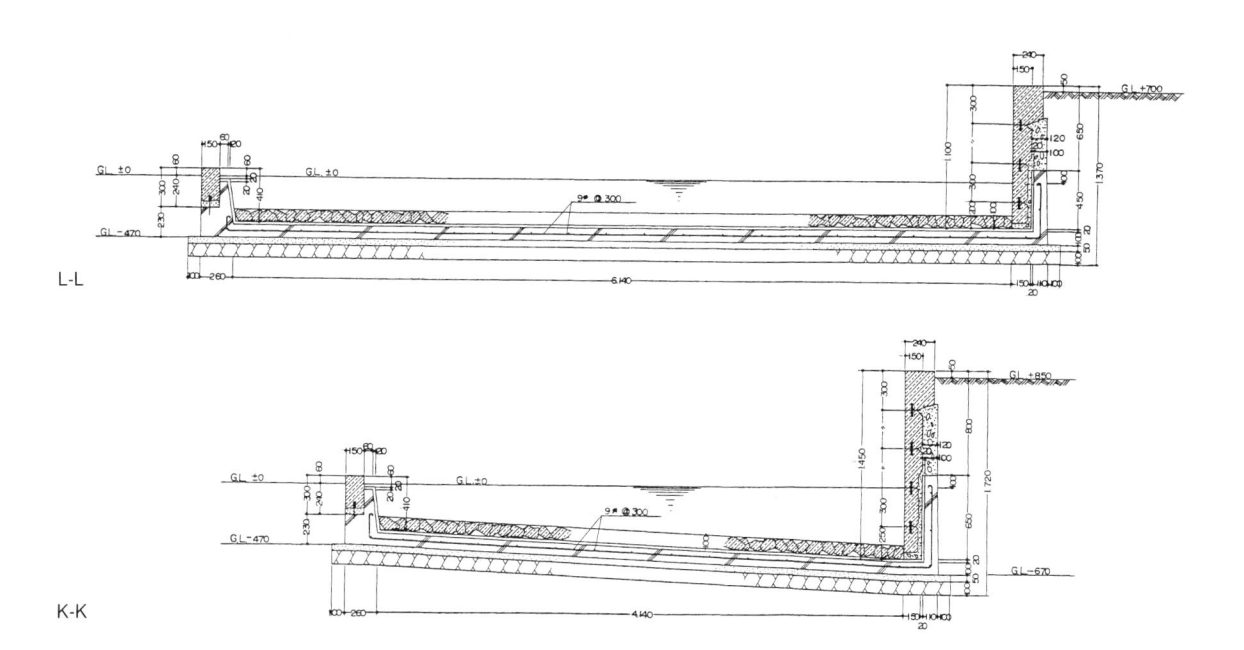

カスケード池部断面詳細図

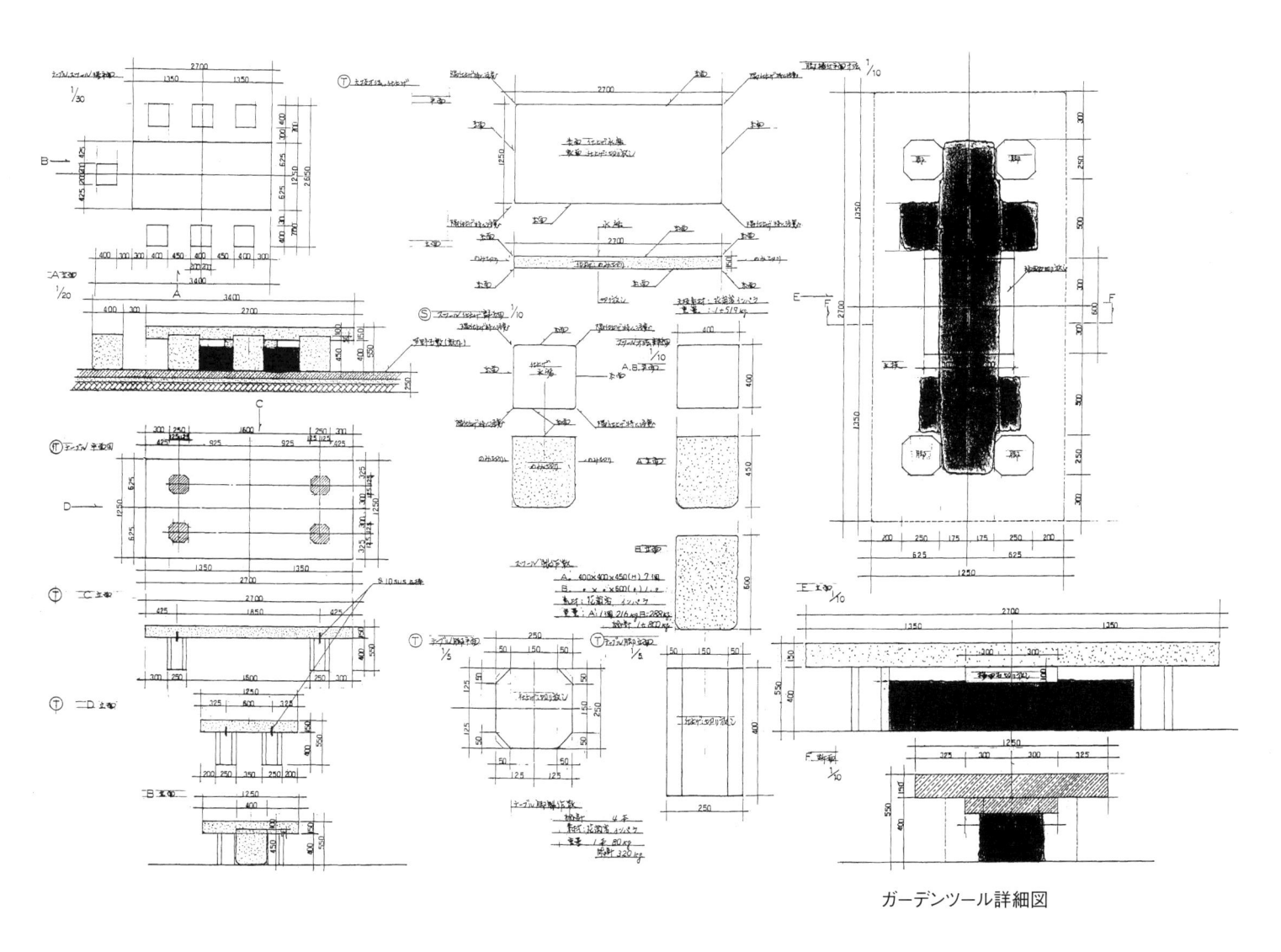

ガーデンツール詳細図

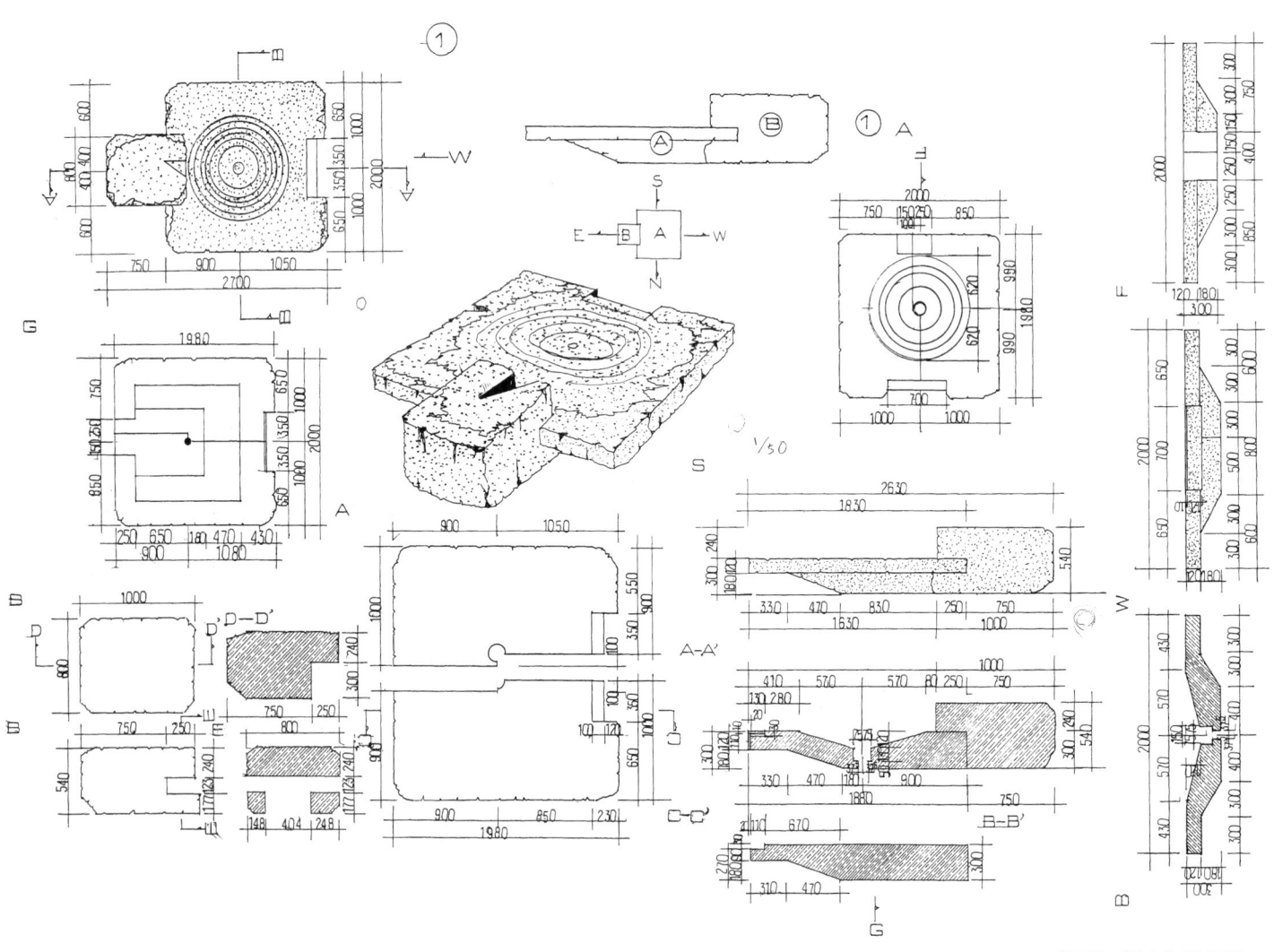

「日野の湧水」詳細図

東銀栗田ビル

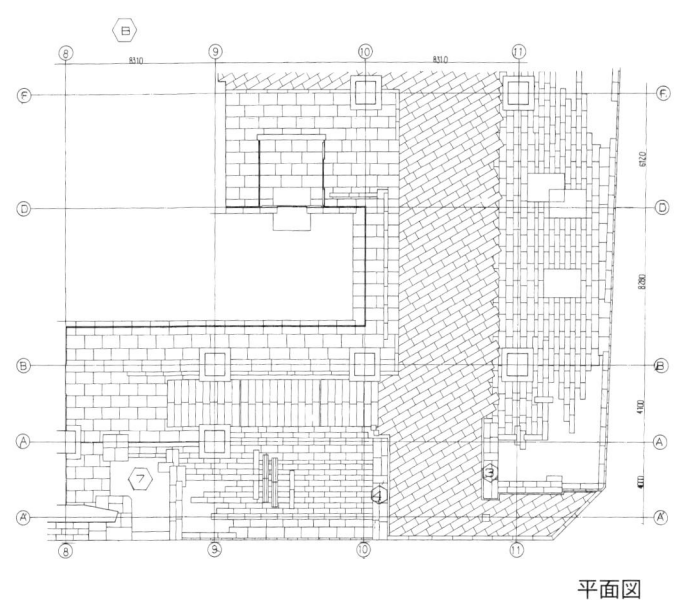

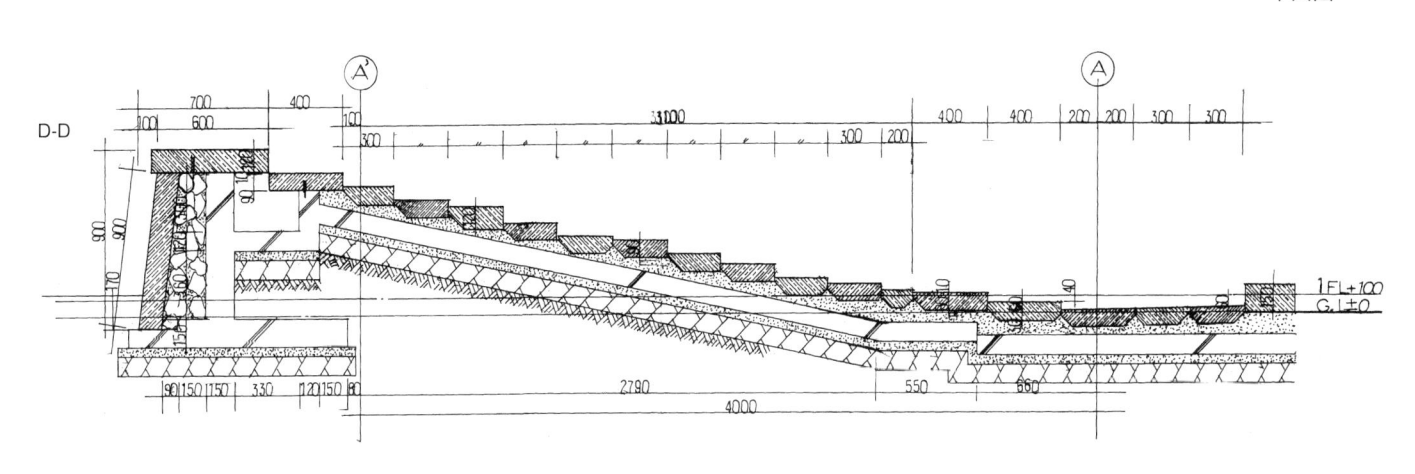

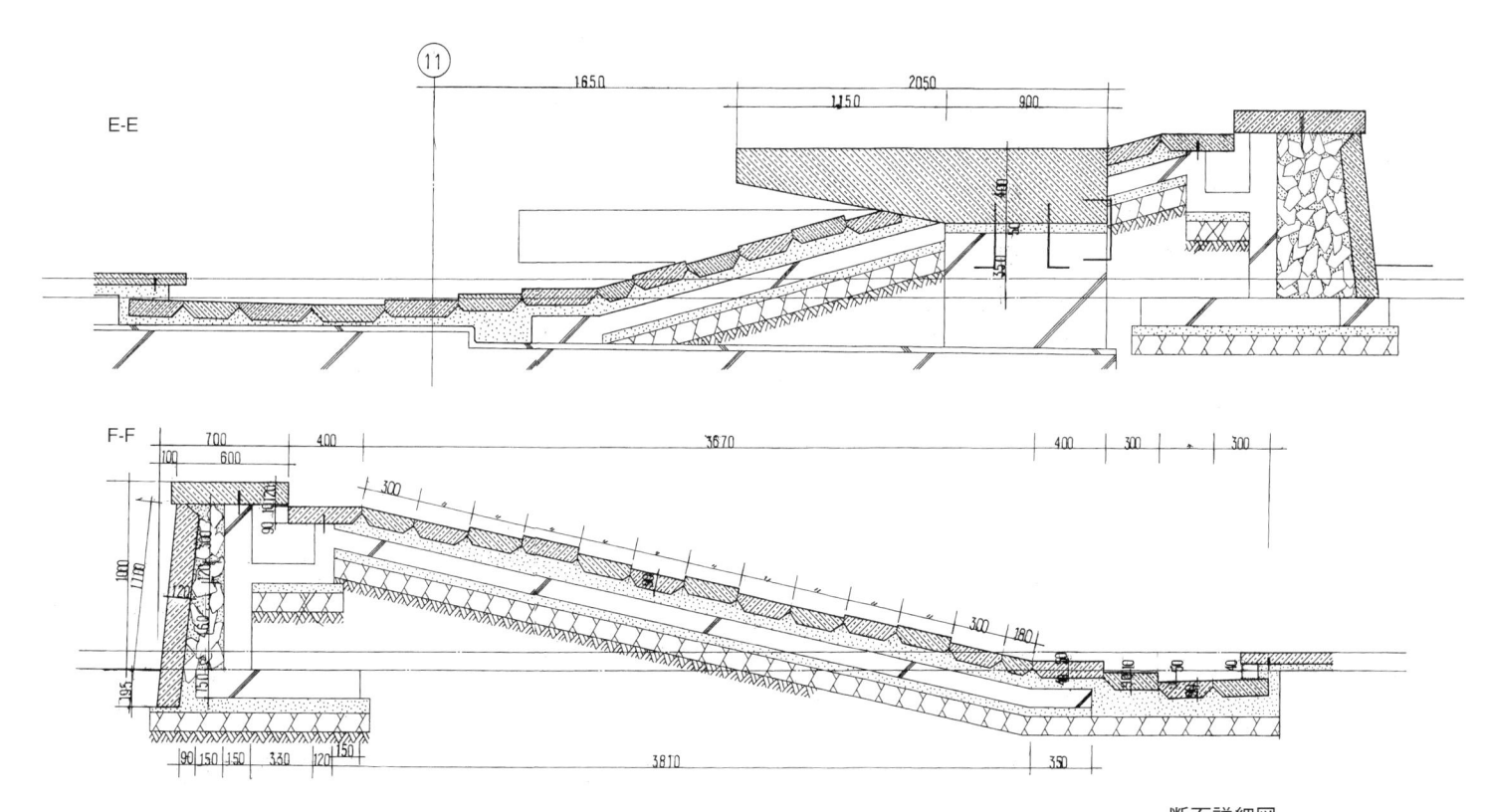

平面図

D-D

E-E

F-F

断面詳細図

①②門柱平面図

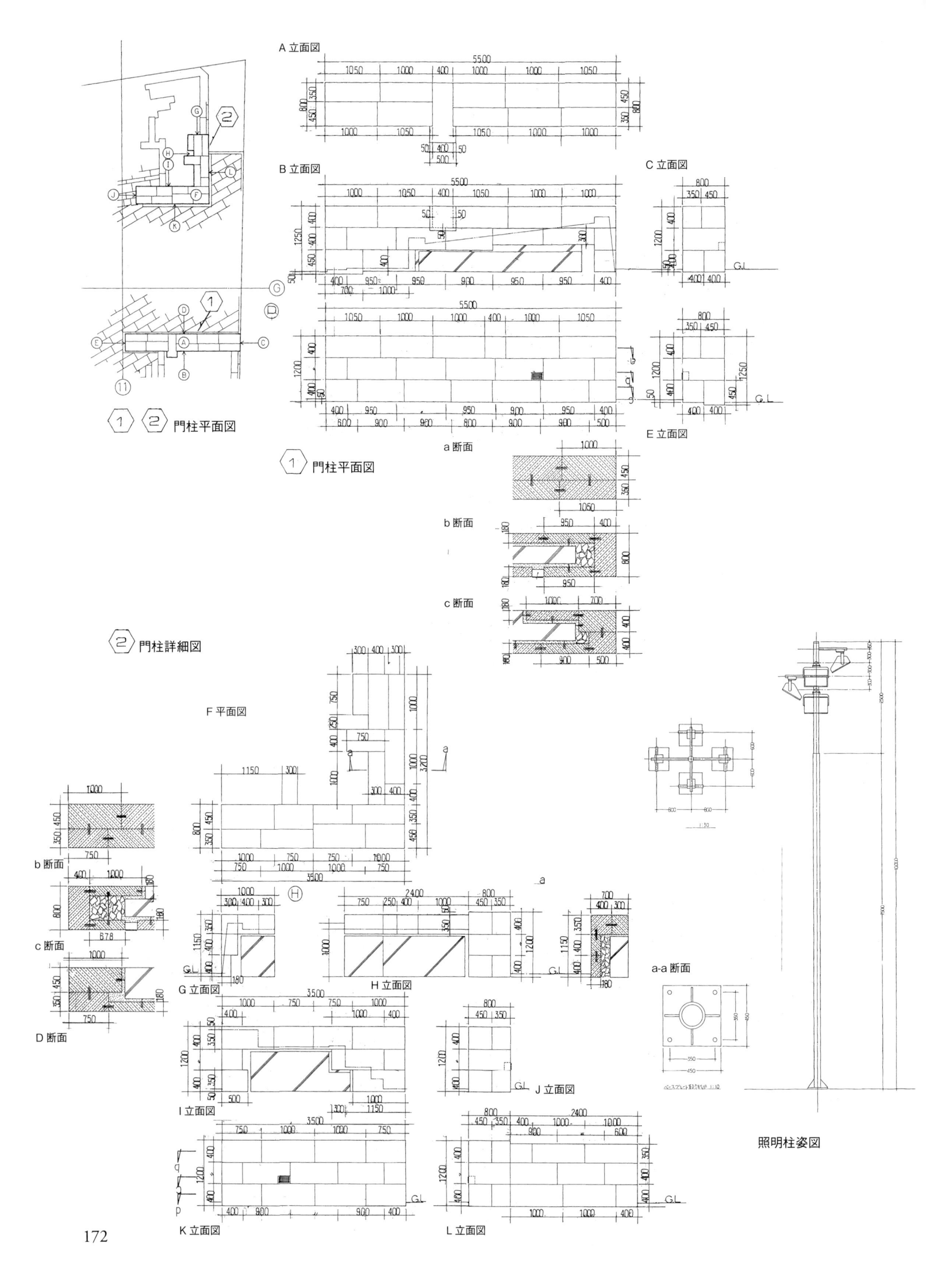

①門柱平面図

②門柱詳細図

A立面図
B立面図
C立面図
E立面図
F平面図
G立面図
H立面図
I立面図
J立面図
K立面図
L立面図

a断面
b断面
c断面
D断面

a-a断面

照明柱姿図

NTT広島仁保ビル

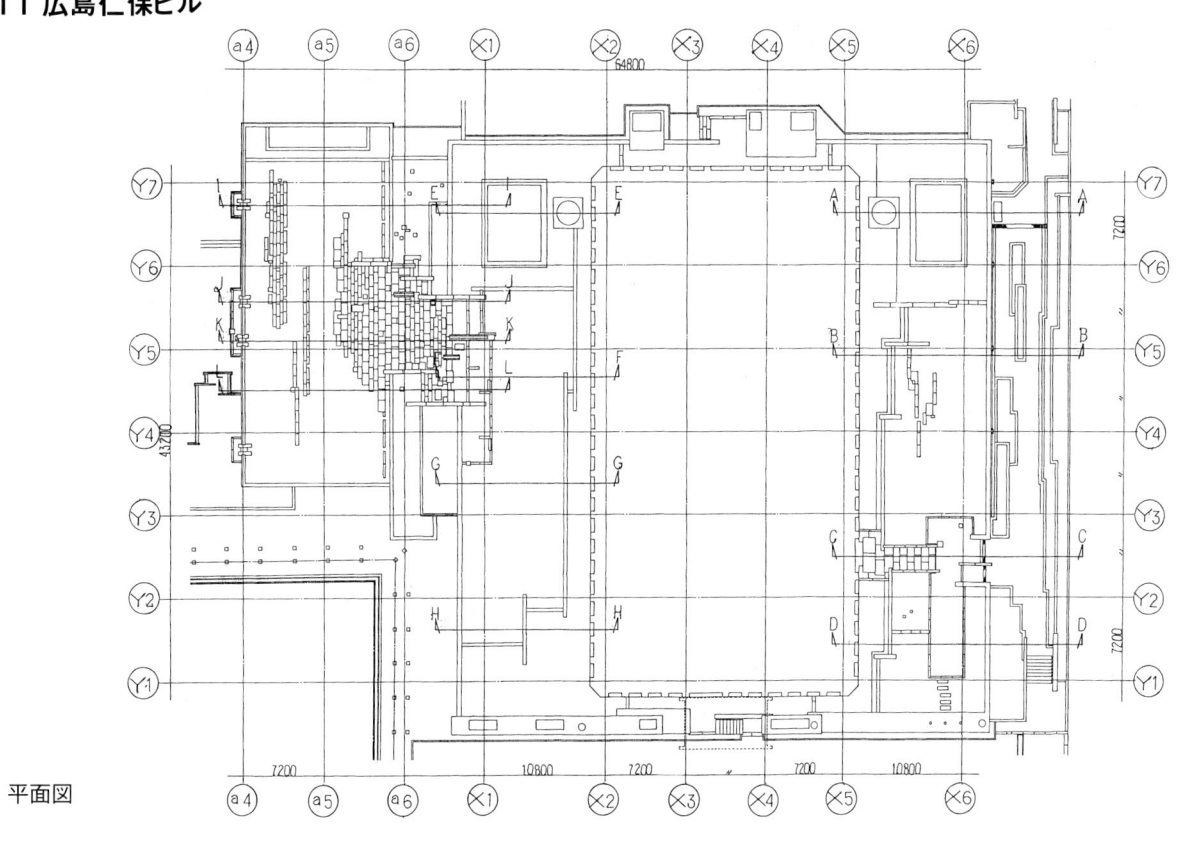

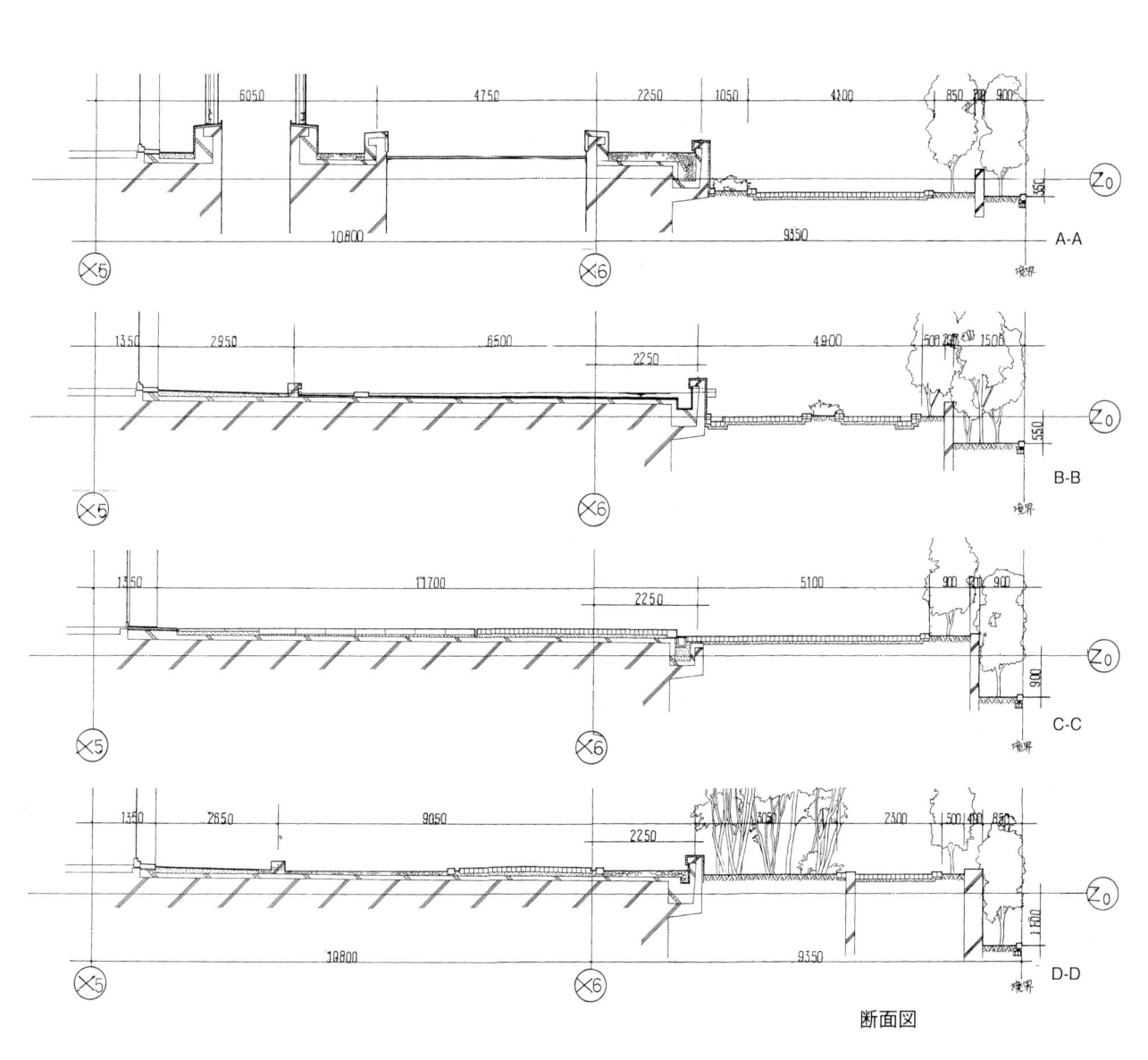

平面図

断面図

174　断面図

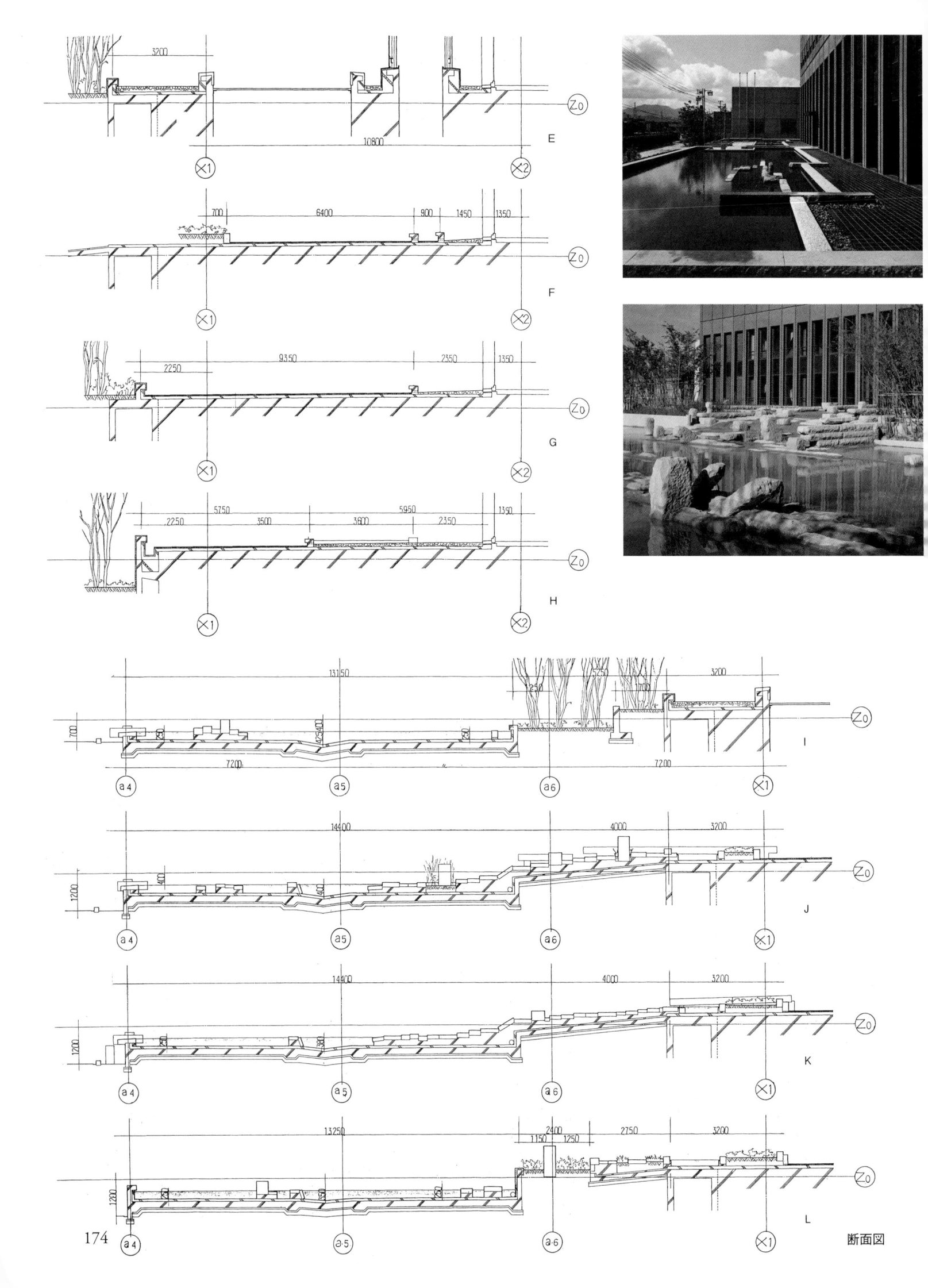

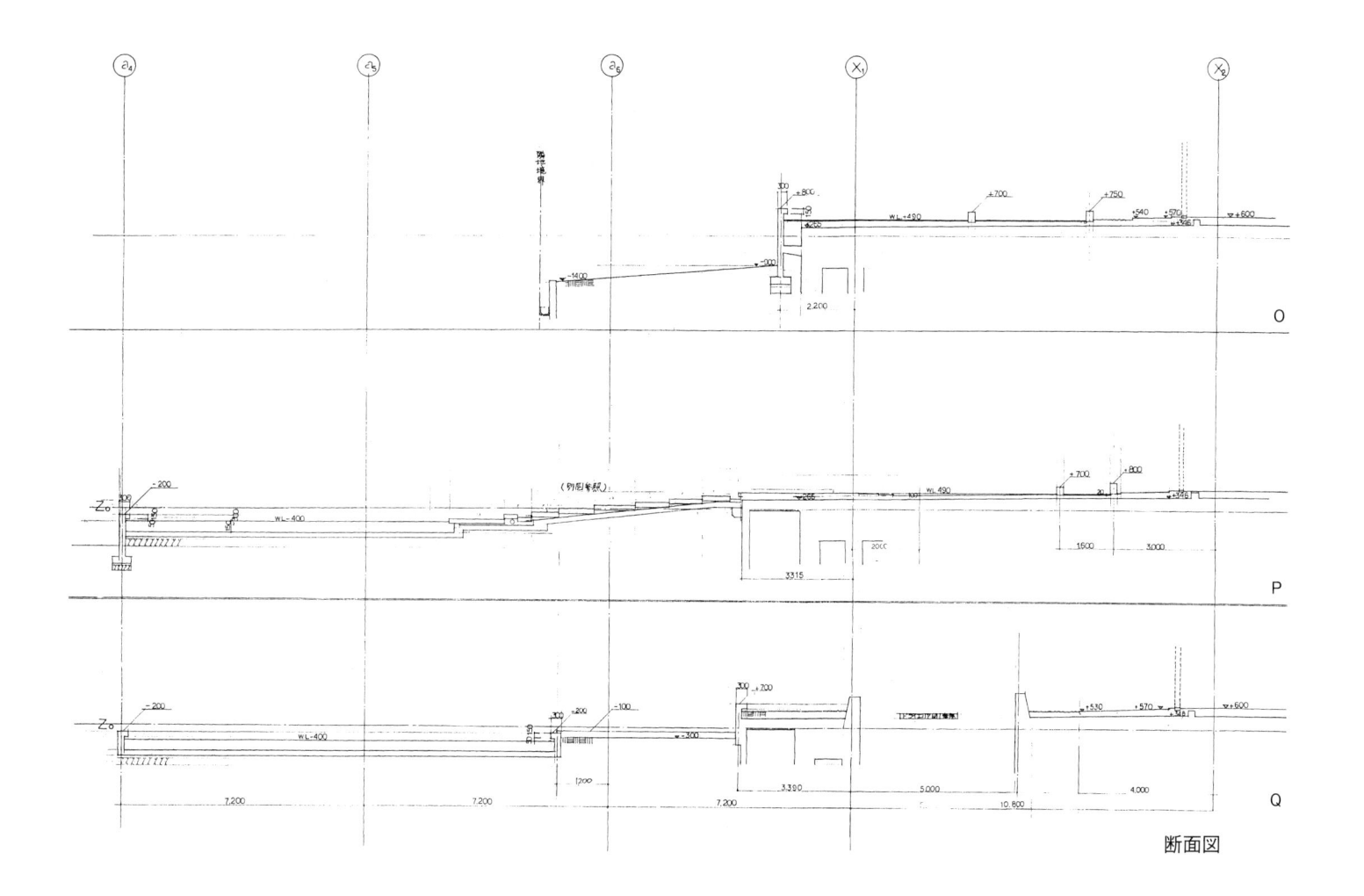

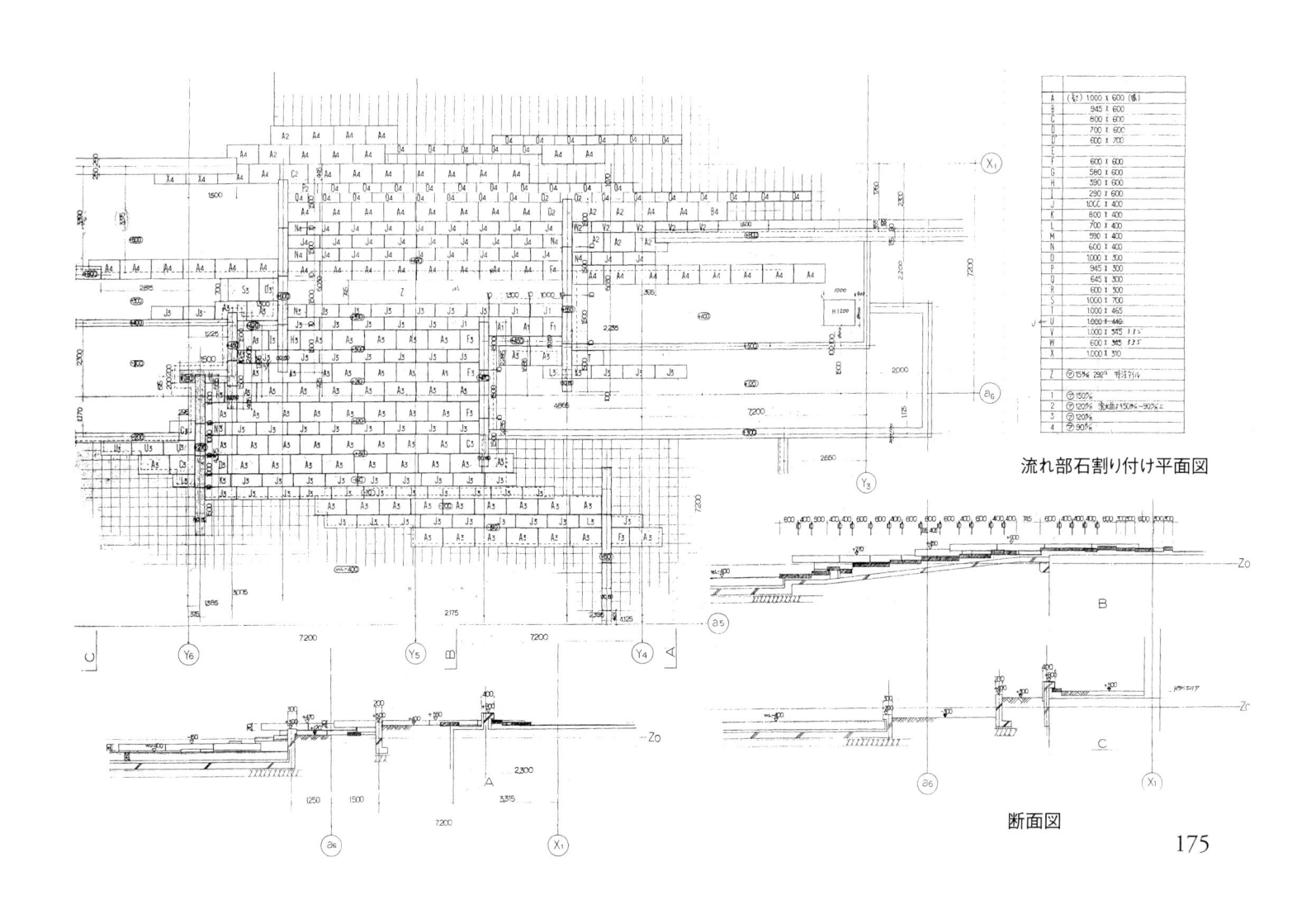

断面図

流れ部石割り付け平面図

断面図

工学院大学八王子校舎 5 号館～11 号館

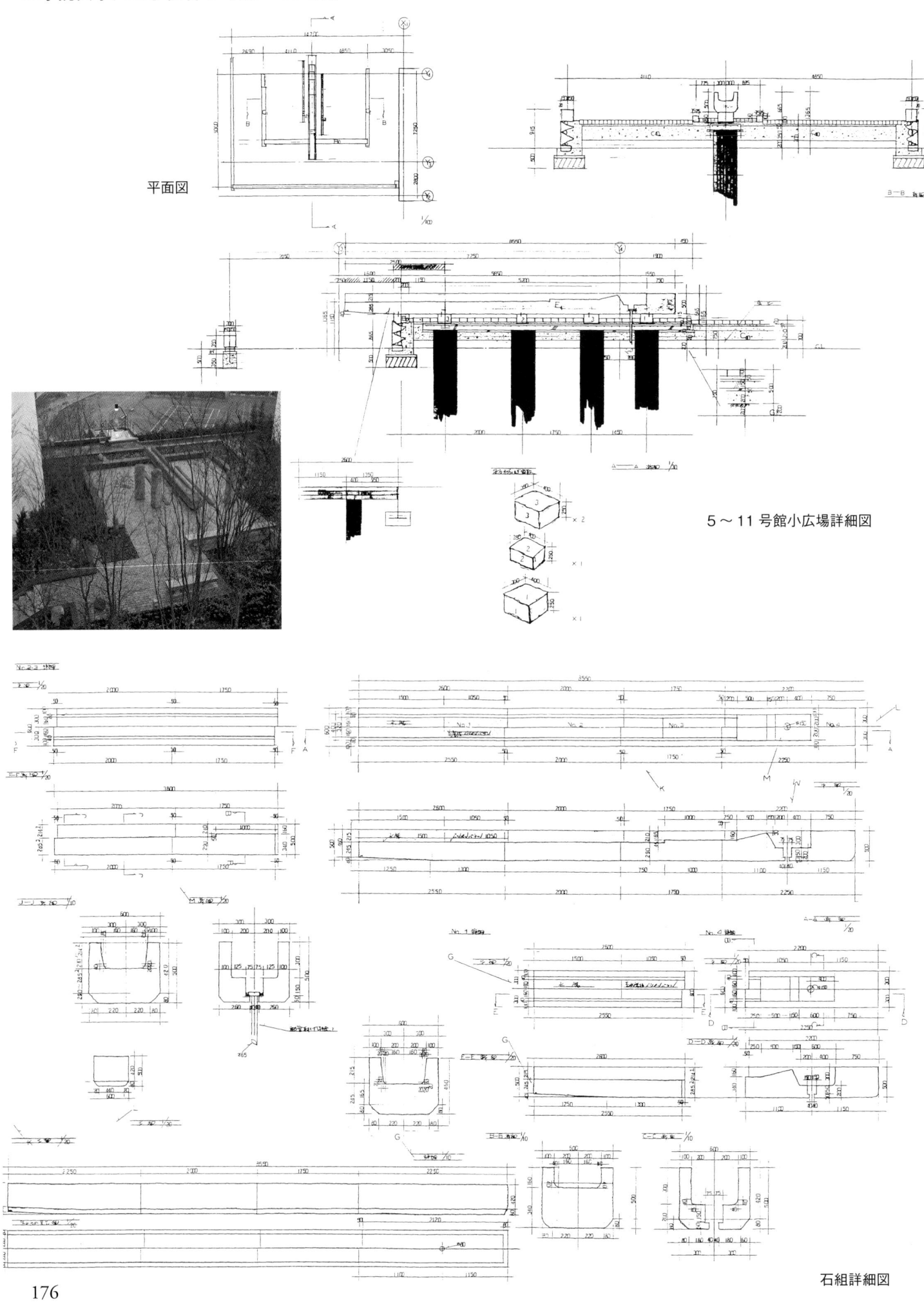

平面図

5～11 号館小広場詳細図

石組詳細図

石組詳細図

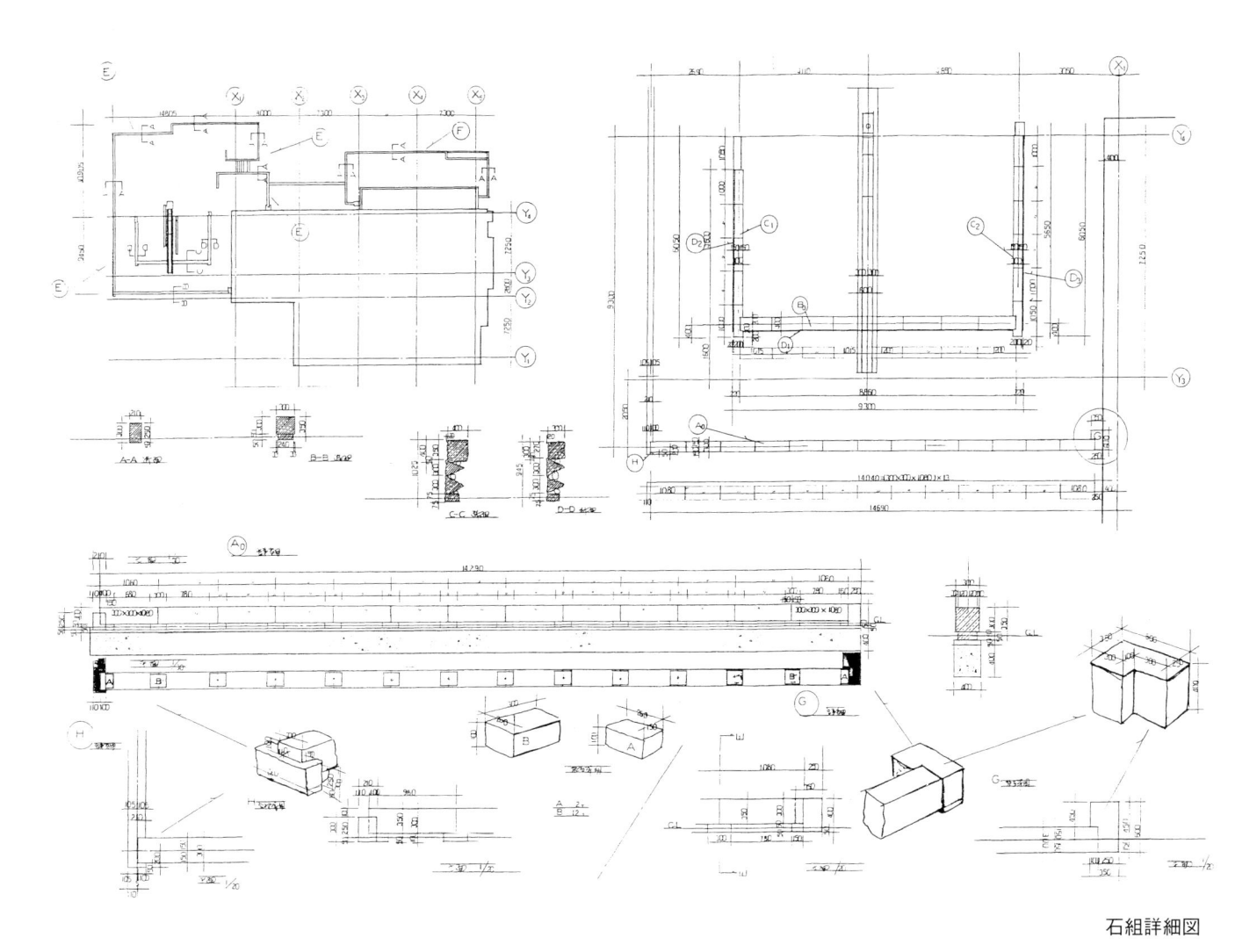

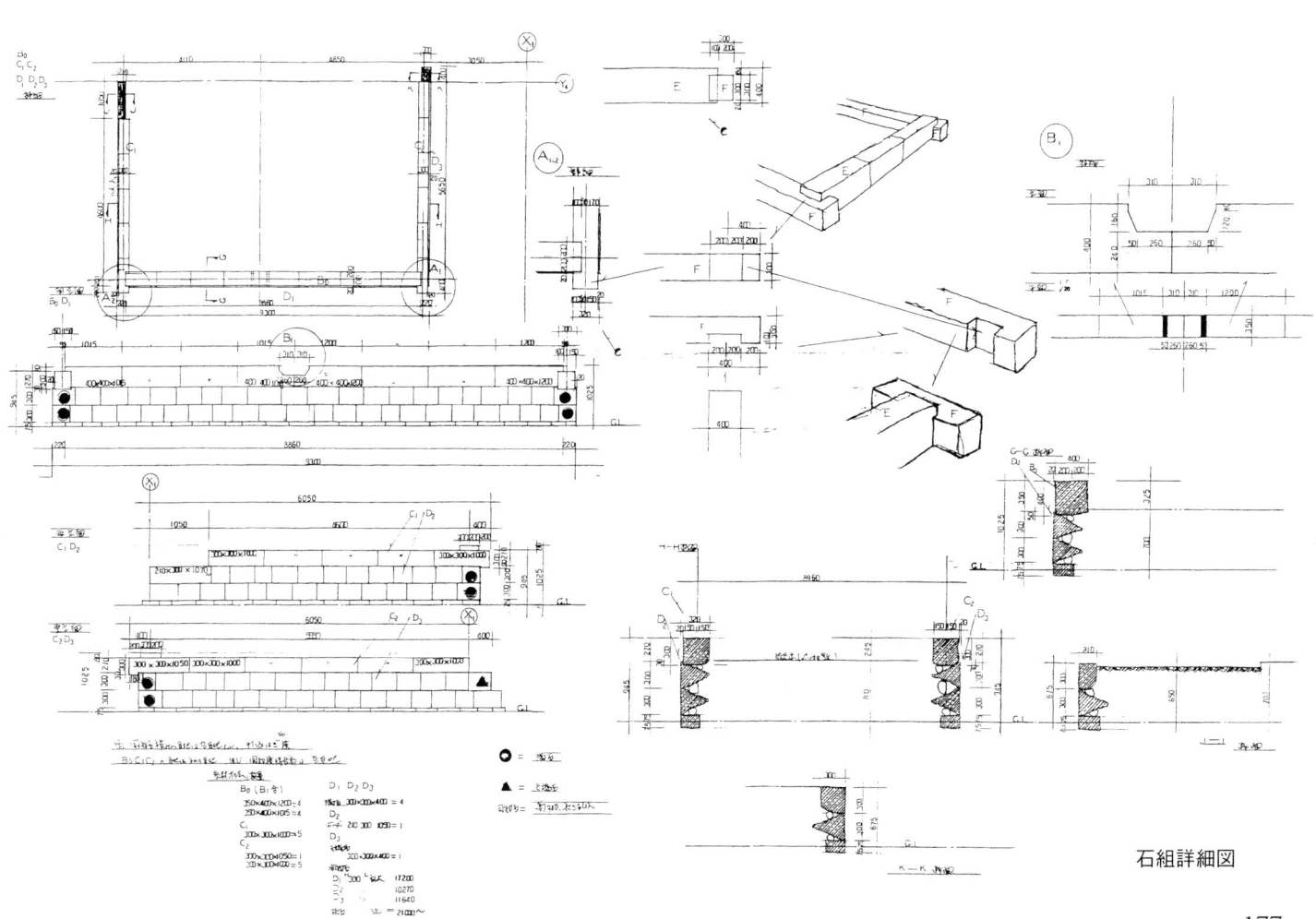

石組詳細図

工学院大学八王子校舎 3 号館

平面図

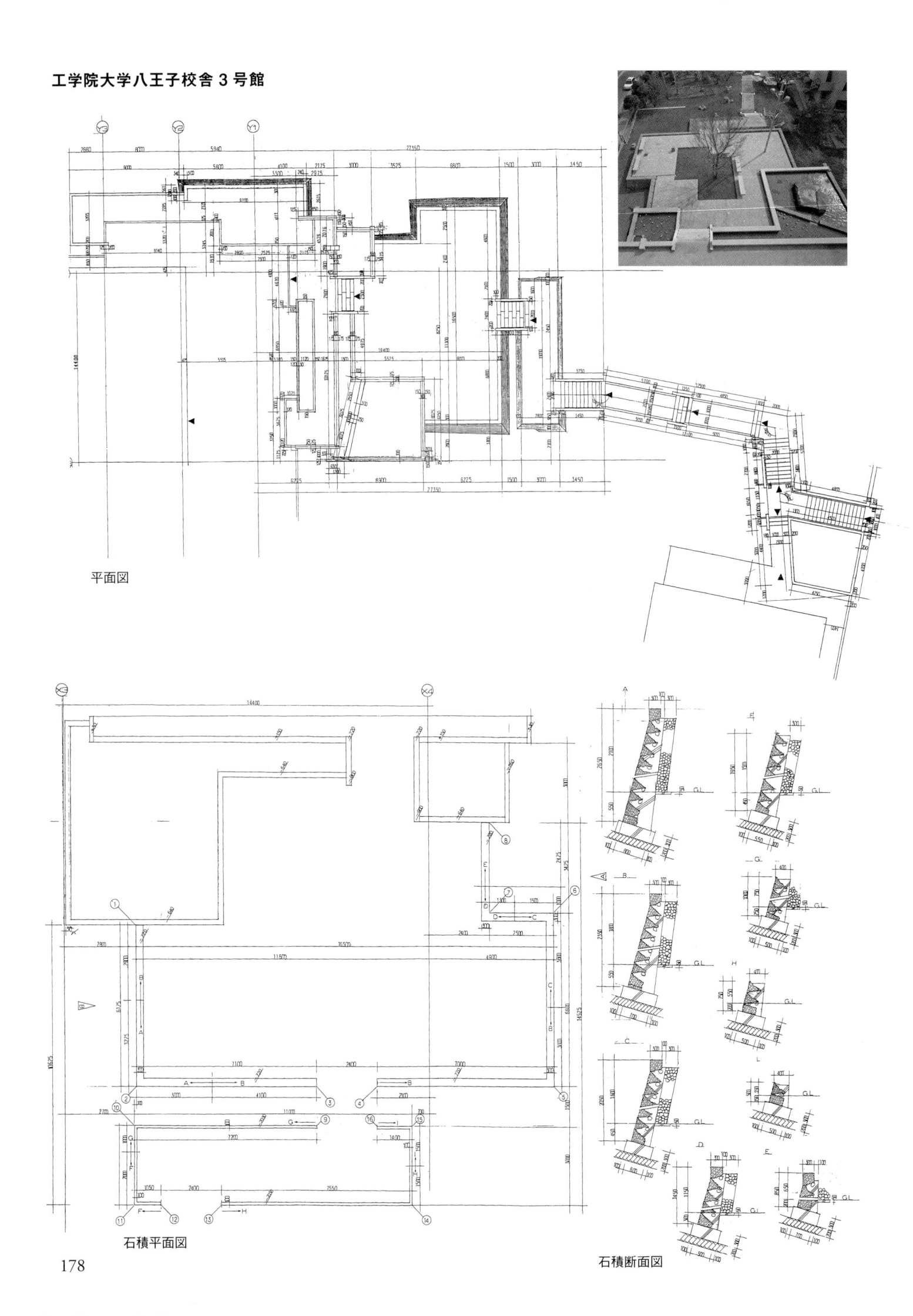

石積平面図

石積断面図

笠石据付平面図

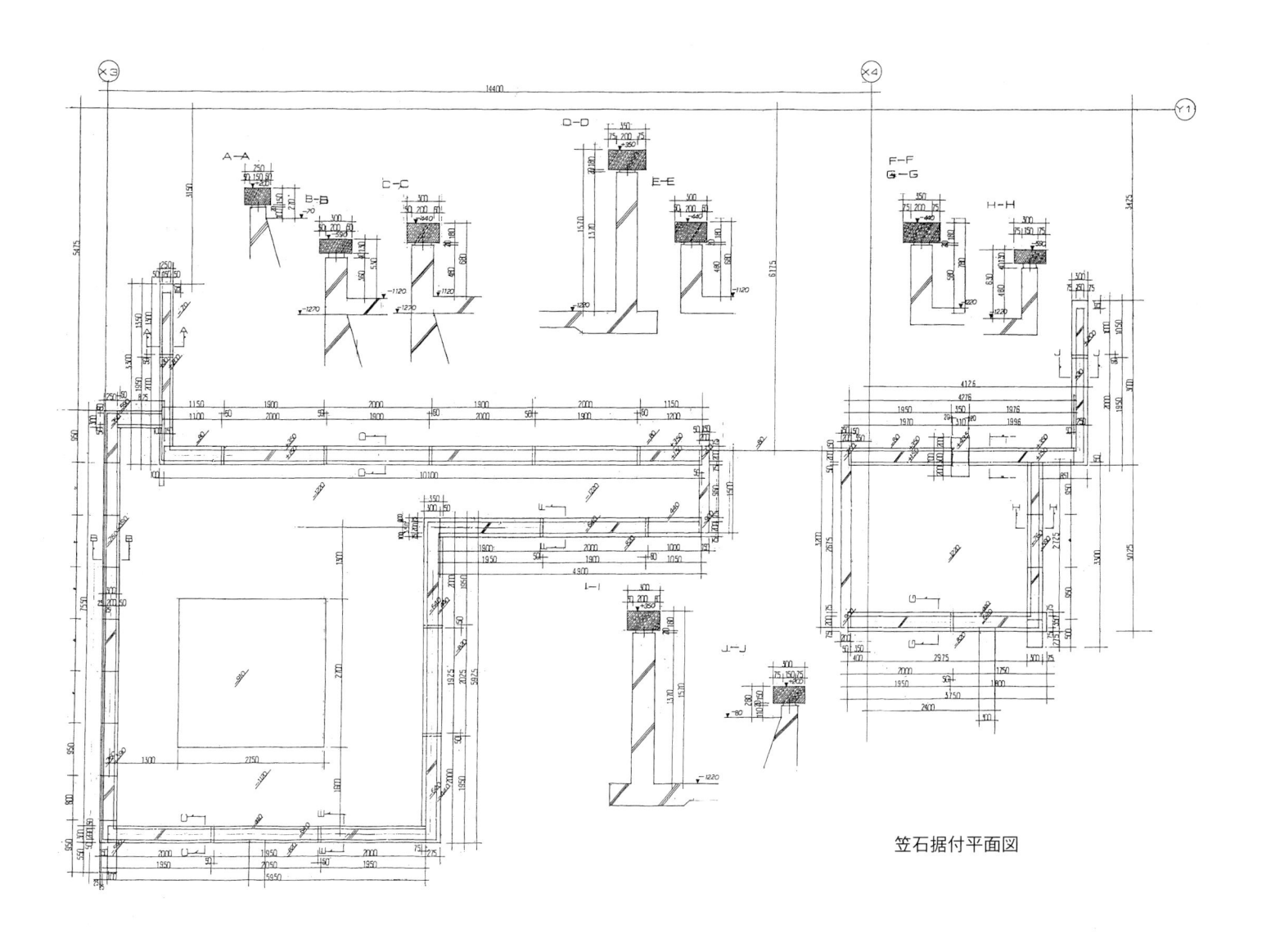

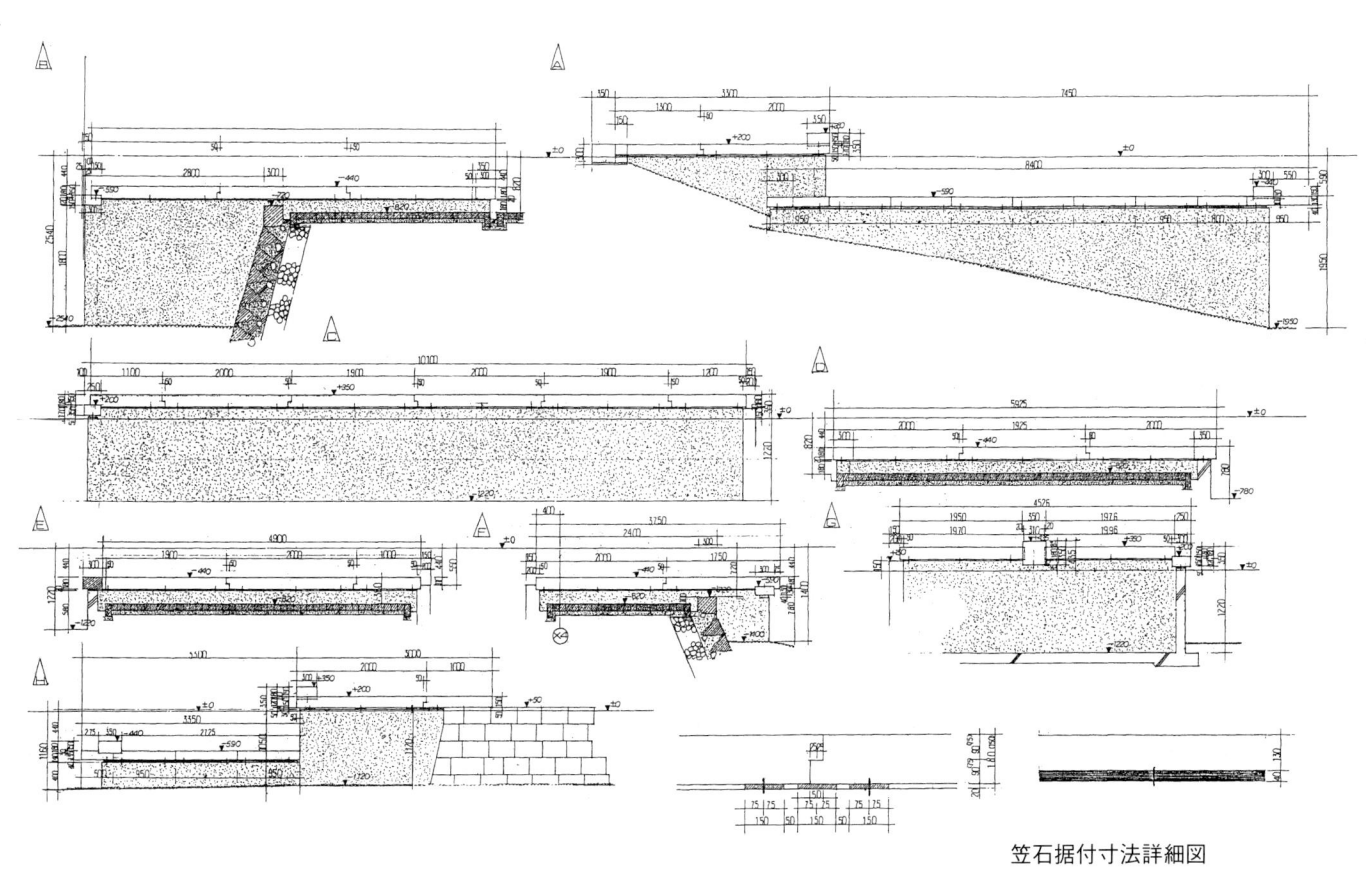

笠石据付寸法詳細図

階段 C, D, E, F, G, H, I 詳細図

湧水・池詳細図

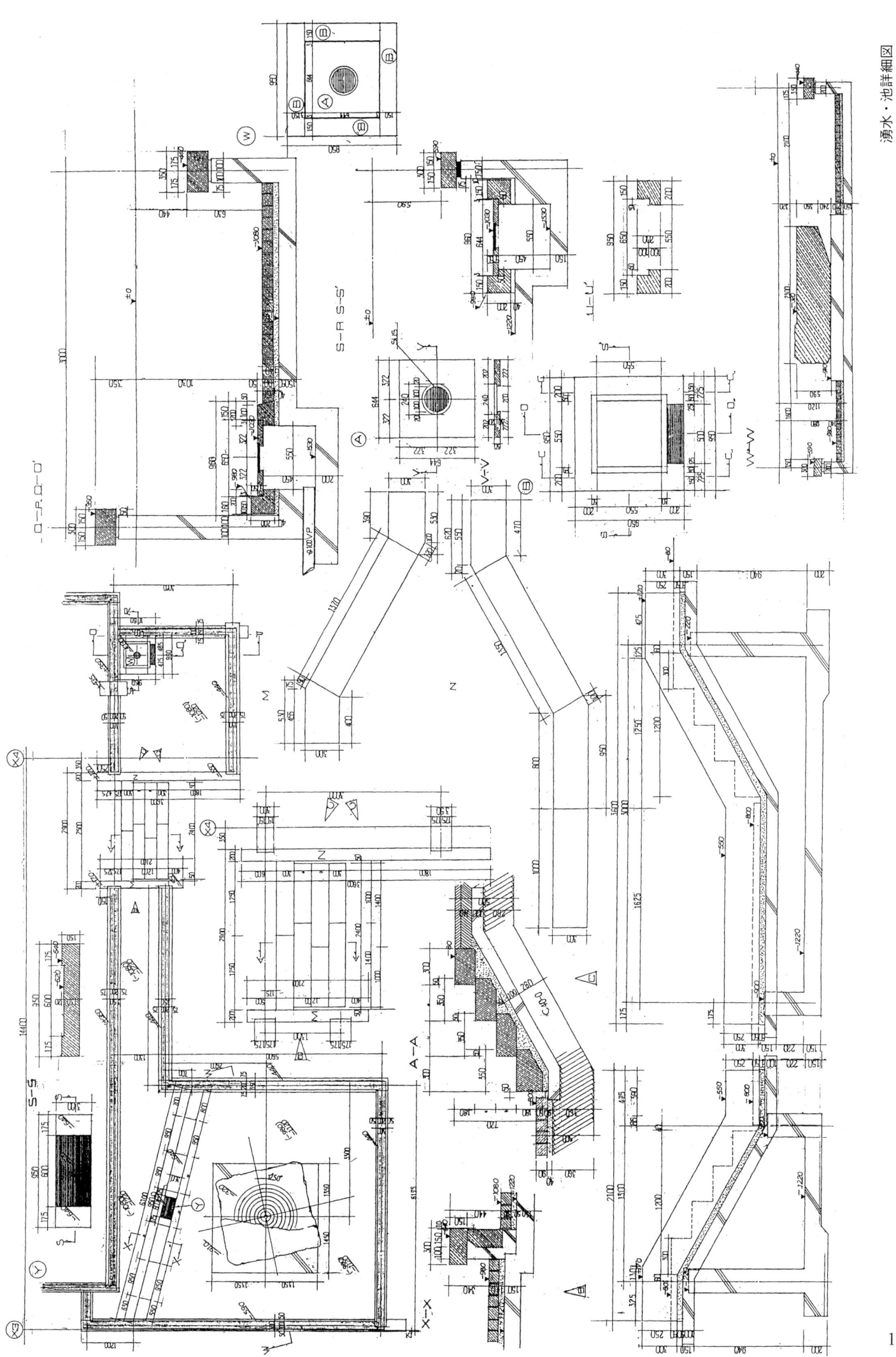

三番町 KSビル

両側石積詳細図

両側階段廻り詳細図

北面石積詳細図

東面基礎躯体平面図・展開図

東面石積など石制作躯体詳細図

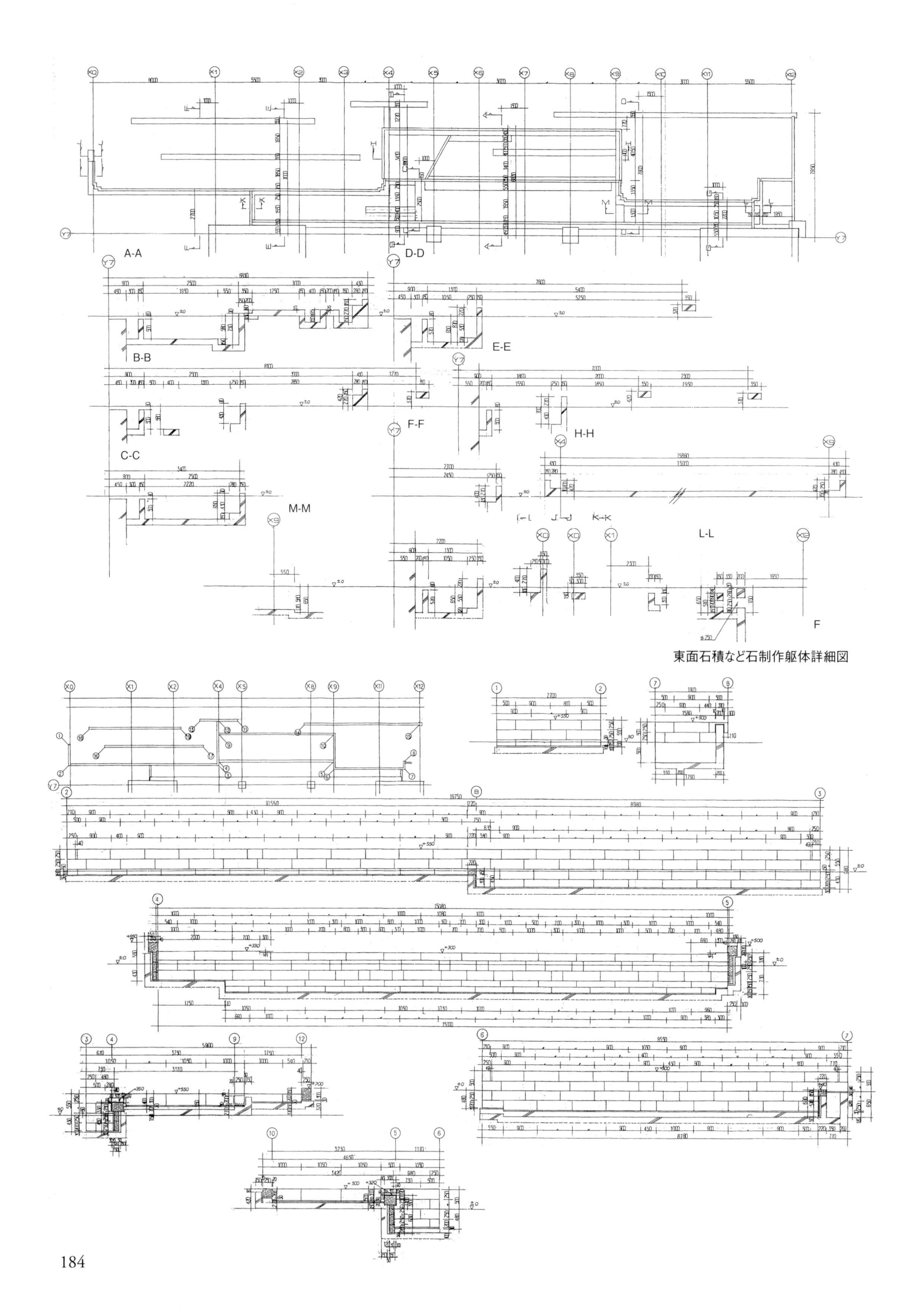

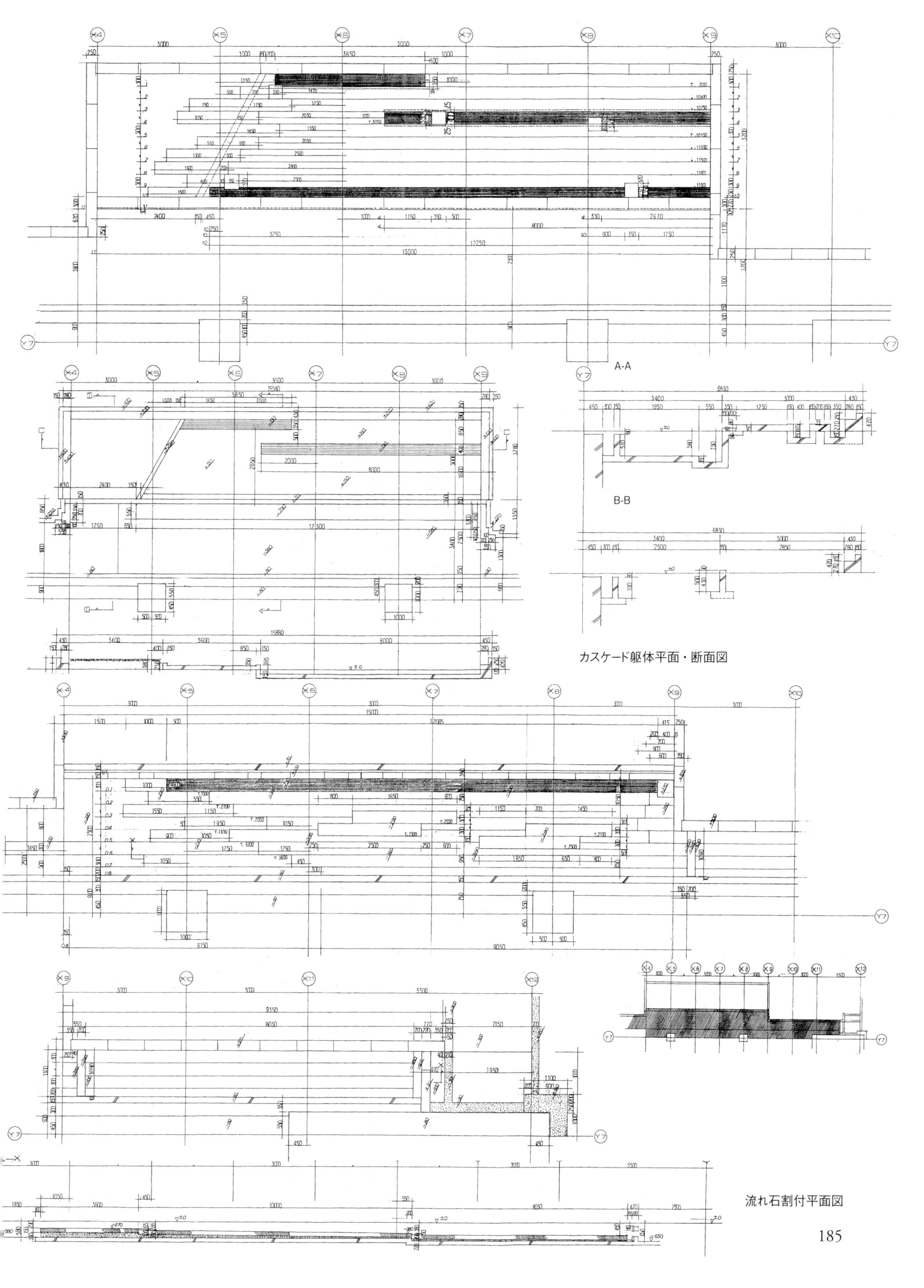

カスケード躯体平面・断面図

流れ石割付平面図

185

東空間石ボーダー土留展開図・断面詳細図

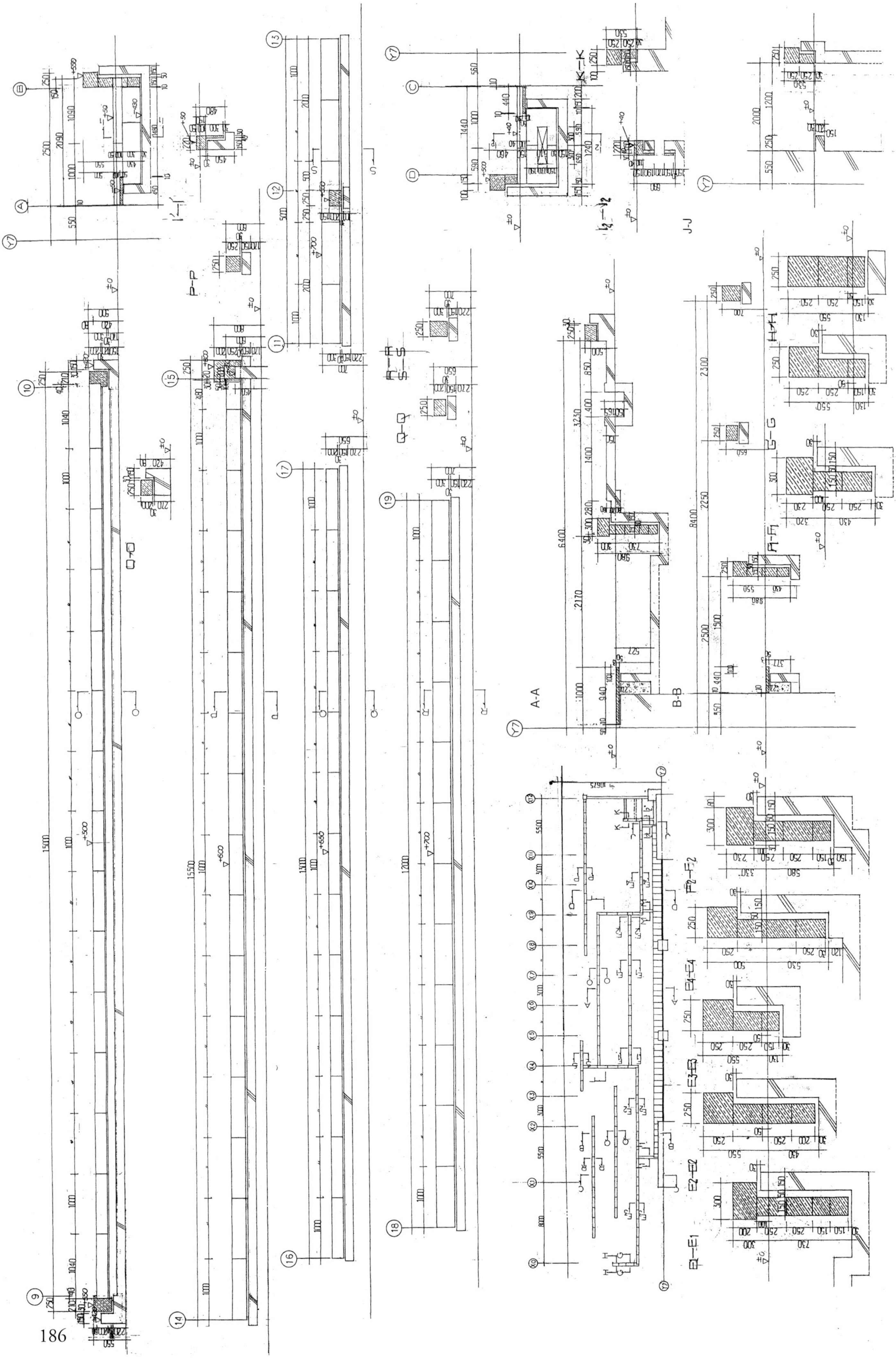

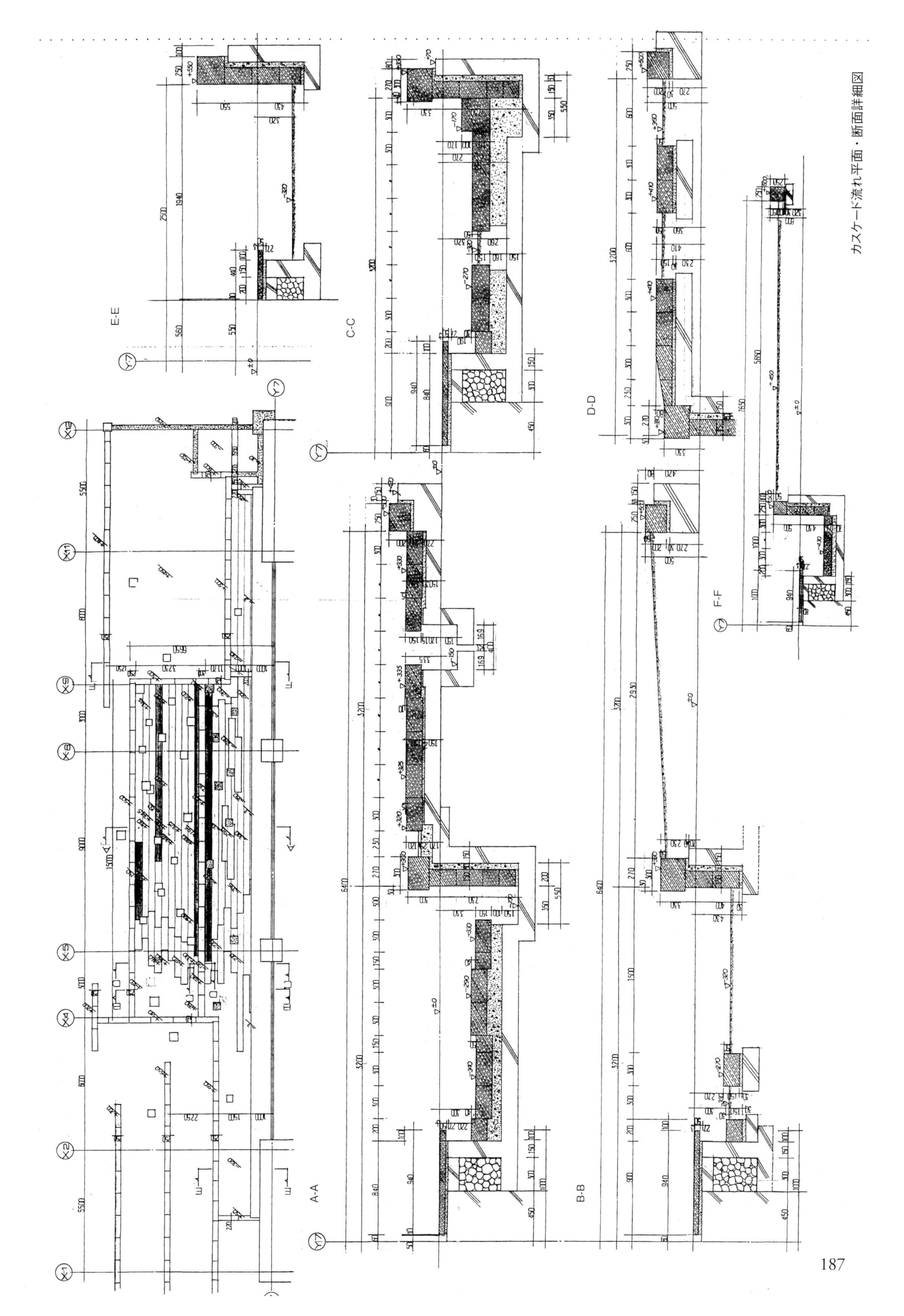

カスケード流れ平面・断面詳細図

データ
Data

京王プラザホテル（北館）4号街路 外空間
1971年
用石の種類
石　積　　　　　　　　新小松石切石
化粧ボーダー　　　　　〃
斜面敷石　　　　　　　芦野石切石
縁石支縁石笠石　　　　〃
一部石積　　　　　　　〃
階段石　　　　　　　　稲田石切石
床　石　　　　　　　　〃

1982年
床敷石　　　　　　　　中国産みかげ石
床飾石　　　　　　　　印度産みかげ石ニューインペ

1983年
石　積　　　　　　　　新小松石間知石
階段袖石　　　　　　　新小松石
階段石　　　　　　　　既存転用稲田石
一部縁石　　　　　　　〃
土　留　　　　　　　　鉄筋コンクリート打放
水の様態　　　　　　　湧水、流れ、滝
自動灌水　　　　　　　スプレーノズル式

常緑広葉樹　小高木　サンゴジュ・サザンカ・ヒサカキ・ヒイラギモクセイ・ネズミモチ
　　　　　　低木　　ハマヒサカキ
落葉広葉樹　高木　　コナラ・イヌブナ・クマシデ・アカシデ・カツラ・シャラノキ
　　　　　　　　　　ヤマボウシ・ミズキ・ヤマザクラ・イヌザクラ・エゴノキ・リョウブ
　　　　　　小高木　ハゼノキ・マンサク
　　　　　　低木　　ニシキギ・ムラサキシキブ・ヤマツツジ・ユキヤナギ・シモツケ・ウツギ
　　　　　　　　　　サイシュウトウハギ・ミヤギノハギ
地被植物　　　　　　ヒメシノ（コクマザサ）・タマリュウ

京王プラザホテル7階屋上
1971年
用石の種類
石　積　　　　　　　　芦野石
化粧ボーダー石　　　　〃
床石、敷石　　　　　　〃
飛　石　　　　　　　　〃
階段石　　　　　　　　〃
水の表態　　　　　　　流れ、滝、堀
自動灌水設備　　　　　スプレーノズル式

常緑広葉樹　小高木　サンゴジュ
　　　　　　低木　　サツキ・ツゲ
落葉広葉樹　低木　　ドウダンツツジ・シモツケ・ウツギ

日野自動車工業本社「日野台の社」外空間
1972年
用石の種類
石　積　　　　　　　　赤城眞石（間知石）、白河石（黒目）
ボーダー石、縁石　　　白河石（黒目）、芦野（黒目）
碑　　　　　　　　　　黒みかげ石ラステンバーグ
自動灌水設備　　　　　ポップアップノズル
水の様態　　　　　　　流れ、滝、堀

1982年
用石の種類追加
記念碑（日野の湧水）　安山岩、新小松石

1989年
用石の種類（改修と追加）
カスケード縁石　　　　黒みかげ石ラステンバーグ
ガーデンテーブル、スツール　〃
堀内植栽榧　　　　　　安山岩、白河石（黒）
林内舗石　　　　　　　花崗岩、稲田石
空気抜蓋　　　　　　　花崗岩、眞壁石

常緑針葉樹　高木　　ヒマラヤスギ・サワラ
常緑広葉樹　高木　　シロダモ・アラカシ・ヤブツバキ・サカキ
　　　　　　小高木　サザンカ・ヒサカキ・ヒイラギモクセイ

落葉広葉樹　高木　　クマシデ・アカシデ・イヌシデ・イヌブナ・コナラ・クヌギ・ケヤキ
　　　　　　　　　　アキニレ・カツラ・コブシ・ヤマザクラ・オオヤマザクラ・シャラノキ
　　　　　　　　　　ヒメシャラ・ヤマボウシ・エゴノキ・ハクウンボク・イロハモミジ
　　　　　　　　　　ヤマモミジ・リョウブ・カントウマユミ
　　　　　　小高木　ウリカエデ・マンサク・ヒサカキ
常緑広葉樹　低木　　ヨウシュ・シャクナゲ・カルミヤ
半落葉広葉樹　低木　アベリア
落葉広葉樹　低木　　ユキヤナギ・ドウダンツツジ
羊歯植物　　　　　　クサソテツ
地被植物　　　　　　コウライシバ・ジャノヒゲ

ホテルグランドパレス 外空間
1972年
用石の種類
石　積　　　　　　　　新小松石切石、芦野石切石
化粧土留ボーダー石　　芦野石切石
化粧土留　　　　　　　鉄筋コンクリート打放しツツキ仕上
水の様態　　　　　　　滝、流れ
自動灌水装置　　　　　スプレーノズル式

常緑広葉樹　高木　　ウバメガシ
　　　　　　低木　　サツキ・ツツジ
落葉広葉樹　高木　　クマシデ・アラカシ

小西酒造東京支店「坪庭」
1974年
用石の種類
単　体　　　　　　　　新小松石
敷　石　　　　　　　　芦野石切石
水の様態　　　　　　　滝、流れ

常緑広葉樹　低木　　セイヨウシャクナゲ
地被植物　　　　　　タマリュウ

東京銀行青山寮 外空間
1975年
用石の種類
単　体　　　　　　　　新小松石
石　積　　　　　　　　芦野石切石、芦野石雑石
ボーダー石、縁石　　　芦野石切石
敷　石　　　　　　　　芦野石枝石
化粧砂利　　　　　　　鞍馬石五郎太

常緑広葉樹　高木　　アラカシ・マテバシイ
　　　　　　小高木　サンゴジュ
落葉広葉樹　高木　　クマシデ・アカシデ・コナラ・エゴノキ・ヤマザクラ・シャラノキ
　　　　　　　　　　ヒメシャラ・ヤマボウシ・ハクウンボク・ヤマモミジ
　　　　　　低木　　シモツケ・アベリア・セイヨウシャクナゲ
地被植物　　　　　　ビンカミノール・タマリュウ

能登「サンかがや」外空間
1976年
用石の種類
単　体　　　　　　　　芦野石雑石
石　積　　　　　　　　芦野石雑石、石動山みかげ自然石
堀（構築材）　　　　　鉄筋コンクリート打放しツツキ仕上
テラス舗床　　　　　　プレコン板
水の様態　　　　　　　滝、流れ、堀

常緑広葉樹　低木　　ツツジ・サツキ
地被植物　　　　　　高麗シバ（コウライシバ）

東銀栗田ビル 外空間
1978年
用石の種類
門柱（石積）　　　　　新小松石切石
石　積　　　　　　　　芦野石切石、芦野石雑石
結果表示石　　　　　　芦野石板石
犬走り敷石　　　　　　稲田石板石
　〃　縁石　　　　　　稲田石切石
客溜外壁　　　　　　　稲田石板石
舗　床1　　　　　　　塩山石ピンコロ
　〃　2
飾　石　　　　　　　　スウェーデンみかげ石エメラルドグリーン
館名石　　　　　　　　小松石
水の様態　　　　　　　流れ

落葉広葉樹	高木	コブシ（1株のみ）・アカシデ（株立1株のみ）
山草類		セキショウ
地被植物		タマリュウ・ビンカミノール

NTT 広島仁保ビル 外空間
1980 年
用石の種類

石　積		芦野石切石
造型貼石		〃 、芦野石板石
ベンチ		芦野石切石
舗床（アプローチ）		みかげ石尾立石板石
		みかげ石北木石ピンコロ
舗床（陶板）別注		
	伊奈製陶	t40×400×200
流れ床（伊奈プレート）		t10×400×400
水の様態		流れ、堀
自動潅水設備		一部スプレーノズル式

常緑広葉樹	高木	ウバメガシ・マテバシイ
	小高木	ヒイラギモクセイ・ネズミモチ
常緑針葉樹	高木	サワラ
	低木	ハイビャクシン
落葉広葉樹	高木	クマシデ・アカシデ・コナラ・ケヤキ・ヤマザクラ・ソメイヨシノ
		ネムノキ・エゴノキ・ヤマモミジ
	低木	ユキヤナギ・シモツケ・ミヤギノハギ

工学院大学八王子校舎5号館〜11号館 外空間
1986 年
用石の種類

石　積		新小松石間知石、芦野石切石
見切石		稲田石切石
床鼻石		〃
階段石		〃
床　貼		稲田石板石
舗　床		塩山石ピンコロ、福島石ピンコロ
立石、石組		芦野石雑石
単　体		新小松石
ベンチ		稲田石
看板石		イタリア産みかげ石
方位石		稲田石
館名石		〃
コンクリート打放し		塗装仕上、ハツリ仕上
舗　床		塩山石ピンコロ、アスコン、アサノ洗出平板
水の様態		流れ、落水

常緑針葉樹	幼木	法面保護	クロマツ
常緑広葉樹	幼木	法面保護	ウバメガシ・アラカシ
落葉広葉樹	幼木	法面保護	アカシデ・ヤシャブシ・コナラ・ヤマボウシ
低木	幼木	法面保護	シモツケ・ユキヤナギ・ウツギ・ヒメウツギ
常緑針葉樹	高木		サワラ
常緑広葉樹	高木		ヤブツバキ
	小高木		ヒサカキ・ヒイラギモクセイ
	低木		セイヨウシャクナゲ
落葉広葉樹	高木		クマシデ・アカシデ・イヌシデ・イヌブナ・エノキ・カツラ・コブシ
			ヤマザクラ・シャラノキ・ヒメシャラ・ヤマボウシ・ヤマモミジ
	小高木		マンサク・ニシキギ・オトコヨウゾメ・ガマズミ・ムラサキシキブ
	低木		ユキヤナギ・シモツケ・ヒメウツギ・ウツギ
地被植物			ヒメシノ（コクマザサ）

工学院大学八王子校舎3号館 外空間
1987 年
用石の種類

石　積		新小松石間知石
〃		コンクリート間知石
コンクリート擁壁笠石		新小松石切石、稲田石切石
コンクリート間知石積一部笠石		芦野石切石
縁　石		中国産みかげ石、芦野石切石
階段石		稲田石
階段袖石		〃
階段石		一部花崗岩、大東石
床鼻石		稲田石
ベンチ		〃
水の様態		湧水、流れ

常緑広葉樹	高木	ヤマモモ
	小高木	ヒサカキ
	低木	セイヨウシャクナゲ

落葉広葉樹	高木	クマシデ・アカシデ・イヌシデ・イヌブナ・カツラ・コブシ
		ヤマザクラ・シャラノキ・ヒメシャラ・ヤマボウシ・エゴノキ
		ハクウンボク・マユミ・アオハダ
	小高木	ウリハダカエデ・ヒサカキ
	低木	シモツケ・ユキヤナギ・ヤマブキ・ウツキ・ガマズミ
		ムラサキシキブ
半落葉広葉樹		アベリア
地被植物		ヒメシノ（コクマザサ）

スコープ本社 GLASS HOUSE 外空間
1991 年
用石の種類

石　積		黒みかげ石ラステンバーグ
縁　石		〃
転　石		〃
床		志野陶石
石積附属壁		コンクリート打放しツツキ仕上
館名石		ラステンバーグ
キャビネット扉		方形型鋼
砂利など		那智黒、石灰砕石

常緑広葉樹	高木	アラカシ
	低木	セイヨウシャクナゲ
落葉広葉樹	高木	ヤマモミジ・イタヤカエデ・ヤマボウシ
地被植物		ヤブラン・ヒメシノ（コクマザサ）

三番町 KS ビル 外空間
1991 年
用石の種類

石積（笠石共）		黒みかげ石ラステンバーグ
カスケード縁石、床		〃
流れ床		〃
転石		〃 、芦野石、中国産みかげ石
化粧土留ボーダー石		〃 、芦野石
舗床板石、ポーチ板石		稲田石、中国産ピンコロ
床鼻石		〃
窓石		〃
階段		〃
擁壁、17 柱		コンクリート打放直仕上（ツツキ）

常緑広葉樹	高木	アラカシ・シラカシ・ヤブツバキ
	小高木	ヒイラギモクセイ・ヒサカキ
	低木	セイヨウシャクナゲ
落葉広葉樹	高木	カツラ・シャラノキ・ヤマボウシ・コブシ・ケヤキ・アカシデ
		クマシデ・イヌシデ・ヤマモミジ
	小高木	ウリカエデ・ナツハゼ
	低木	ムラサキシキブ・ガマズミ・ニシキギ・ヤマブキ・ヤマツツジ
		ユキヤナギ・シモツケ・ドウダンツツジ
山草類		ヒガンバナ・キキョウ・カタクリ・イカリソウ・ノカンゾウ・キスゲ
		エビネ・セキショウ
地被植物		ヒメシノ（コクマザサ）・タマリュウ

エピローグ
Epilogue

　本文に述べたように、都市の外部空間を表現空間に選び、自身に規律を課し、素材と造形手法を厳しく限定し、一定の空間体験に近づこうと、迫力ある空間の創成を目指し20数年の時を重ねて来た。その20数年間に目指した空間が、結果として、1972年の京王プラザホテル4号街路外空間の完成をはじめとして、1991年に完成した三番町ビル外空間に終わった5つの外空間が、当初の目的に達し得たように思われる。

　上述の5つの空間の外に敷地の内部に取り込まれている空間も併せて掲載した。それらの空間は、5つの空間を創成した道程での試行錯誤を含む多くの手掛かりを生んでいる。そういうことで、それら数々の空間を参考までに載せた。それぞれは、読者が5つの空間創成への理解をより深めるであろうことを期待したからである。

　迫力ある空間の創成は、その構成に当たって、「自然な自然」と「意図され逞しい自然」とが組み合ったものがよかろうとしたが、どうやら私は創ろうとしたそれぞれの都市空間に生んだ「新しい自然」とそれぞれの空間に介在する人々との間に対話の生まれることを願っていたのかもしれない。

　現代のような時間を短縮して結果を求める時代にあって、価値ある外部空間とは何か、もう一度問い直す時が真剣に問われてよいのではなかろうか。

As described in the main text, I have been impose discipline on myself to design urban exterior spaces, working within the constraints of the materials and the construction techniques, trying to get close to a certain spatial experience to build fabulous spaces for more than two decade. As a result, the five *'Gai-Kukan'* projects from the *'Gai-Kukan'* of the Keio Plaza Hotel, Street No.4 in 1972 to that of the Sanbancho Building in 1991 reached close to my original intentions.

In addition to the above five spaces, I introduced some others that built within the confined spaces of a building compound. Those spaces gave me many hints and clues for the designing space for process of trial and error, including of those five spaces. So I expect readers to understand the design process in the five spaces more better.

For to create a powerful space, I preferred a combined 'natural nature' and 'intentional, bold nature', but it seems that I hoped to let a conversation start between the people who living the space and the 'new nature' spaces in which I created, each other.

Nowadays, we are living in the world to which demand do efficient and obtain good results, and therefore, I suppose to rethink about value of *Gai-Kukan* seriously.

1973年(48才)

1986年(60才)

1993年(68才)1月23日の講和会にて

プロフィール
Profile

深谷光軌 (ふかや こうき)

1926 年東京上野東叡山寛永寺、子院等覚院の長男に生まれる。その後、得度僧籍に入り、住職を拒否。62 年より本格的に作庭を開始。現代人の乾いた気持ちに割り込む外空間を創出し続けるが 1997 年 12 月に逝去。享年 71 歳。代表作として日野自動車本社、京王プラザホテル、ホテル・グランドパレス、小西酒造東京支社、東京銀行栗田ビル、広島仁保ビル、工学院大学八王子校舎、三番町 KS ビルなどがある。

Koki Fukaya

Born in 1926 as the eldest son of the Togakuin Temple branch of Toeizan-Kaneiji, Ueno, Tokyo. He entered the Buddhist priesthood but later refused a post of chief priest. In 1962, he started landscape design by proffessional. He continued to create many 'Gai-Kukan' that gently break in on the parched hearts of contemporary people, but passed away at 71 years old in December 1997. His masterpieces are Hino Motors Main office, Keio Plaza Hotel, Hotel Grand Palace, Konishi Brewery Tokyo Branch, Togin Kurita Building, NTT Hiroshima Niho Building, Kogakuin University Hachioji Campus Buildings, Sanbancho KS Building, and others.

■ 1962 年以降の主な設計、制作の実績

年	実績
1962 年	東京タワーボーリングセンター
〃	日本放送協会軽井沢公館
1962〜63 年	八十二銀行長野研修所
〃	〃 迎賓館夜宋荘
1964 年	本田技研工業ウェルカムポイント
1965 年	木村屋総本社三芳工場
〃	ポーラ化成工中央研究所
〃	三和銀行国分寺寮
〃	メキシコ大使館施設
1966 年	麻布レジデンショナルホテル
〃	ホテル宗藤
〃	白洋社本社工場
〃	国立福祉会館
1967 年	山陰合同銀行東京アパート
〃	ニュートーキョー　小田急新館店 "貴船"
〃	明治製菓百合ヶ丘総合グランド
〃	姫の湯ホテル計画
1968 年	朝日生命成人病研究所
〃	〃　本社 5F
〃	ホテル水明館計画
1970 年	第一生命住宅小枝ビル新館
1971〜72 年	日野自動車工業本社
1971 年	京王プラザホテル　　4 号街路
〃	7F 屋上
1972 年	ホテルグランドパレス
〃	松方邸
1973 年	小西酒造東京支店
〃	王子スポーツガーデン
1974 年	レストランあかべこ
〃	小西酒造東京本社
〃	ホテル白水計画
〃	霧島ホテル計画
1975 年	東京銀行青山寮
〃	栗田出版販売本社
1976 年	サンかがや
〃	藤井荘
〃	小寺邸
1977 年	厚別商業センター
〃	イラン Mr ハダザデゲストハウス
1978 年	東銀栗田ビル
1979 年	越ヶ谷市民会館（碑）
〃	島村融氏（墓）
1980 年	広島仁保電電ビル
1982 年	京王プラザホテル 4 号改修
〃	日野自動車本社改修
1983 年	山村家（墓）
〃	京王プラザホテル 4 号街路改修
1984 年	日野自動車本社改修
1986 年	イトーヨーカドーグループ多摩研修センター
〃	工学院大八王子校舎　5-11 号館
1987 年	〃　　　　　　　　　3 号館
1987〜89 年	〃　　　　　　　　　3,5-11 号館
1987 年	水野邸
1990〜91 年	㈱スコープ本社 "グラスハウス"
1991 年	三番町 KS ビル

Editorial and Design

Editorial and Art direction
丸茂　喬
Takashi Marumo

Editorial and Design
丸茂弘之
Hiroyuki Marumo

Cover Design
村上　和
Kazu Murakami

Translation

田辺裕子
Yuko Tanabe

マーク・ピーター・キーン
Marc Peter Keane

Translation Cooperate
一ノ瀬健介
Kensuke Ichinose

参考文献

内田芳明著	風景とは何か／朝日選書445	
中野孝次著	清貧の思想／草思社	
アブラアム・A モール著	叢書 ウニベルシタス363／法政大出版局	
エリザベート・ロメル著	生きものの迷路	
吉田幸男訳		
神崎宣武著	湿気の日本文化／日経新聞	
木村尚三郎著	文明が漂うとき／日経新聞	
マーティン・ガードナー著	"自然界における左と右"／紀伊国屋書店	
坪井忠一、藤井照彦		
小島 弘訳		
司馬遼太郎著	草原の記／新潮社	
司馬遼太郎著	風塵抄／中央公論社	
中野 肇著	"空間と人間 文明と生活の底にあるもの"／中公新書	
河合隼雄著	こころの処方箋／新潮社	
饗庭孝男著	ヨーロッパの四季Ⅰ／東京書籍	
	ヨーロッパの四季Ⅱ	
北尾邦伸著	森林環境と流域社会／雄山閣出版	

Photographers

相原　功 Isao Aihara	表紙カバー、表紙、p20〜25、p47、p68、p72、p73、p74〜p82、p86〜p90、p92、p97、p100〜p105、p154
大橋治三 Haruzou Ohashi	p54、p55、p57 上、p69〜p71、p156、p158〜p160
川澄明男 Akio Kawazumi	p83
信原　修 Osamu Nobuhara	p60
廣田治雄 Haruo Hirota	裏表紙カバー、P5、p36〜p38、p39、p40、p41、p44〜p46、p48、p49、p52、p53、p106〜p109、p111、p113、p115〜p119、p121、p123〜p127、p134〜p136 上、p137、p139、p143〜p150
深谷光軌 Koki Fukaya	p6〜p11、p29、p30
牧　直視 Naoshi Maki	p31 左下、p56
村井　修 Osamu Murai	p28
門馬金昭 Kaneaki Monma	p31 上、右下
SCOPE	P128〜p130、p132

「外空間」を創る

2011年8月15日　第1版発行
発行人：丸茂　喬
編著：深谷光軌
編集協力：倉橋潤吉　山口詳二朗
発行所：株式会社マルモ出版
〒150-0042 東京都渋谷区宇田川町2-1
渋谷ホームズ1405
電話 03-3496-7046
http://www.marumo-p.co.jp
印刷・製本：株式会社ローヤル企画

©Marumo Publishing Co.,Ltd.
Printed in Japan

Inner-Landscape

First Printing, August 2011
Publisher：Takashi Marumo
Written and edited：Koki Fukaya
Editorial Cooperation：Jyunkichi Kurahashi, Shoujirou Yamaguchi
Publishing: Marumo Pubulishing Co.,Ltd
1405 Shibuya-Homes 2-1 Udagawa-chou Shibuya-ku ,
Tokyo 150-0042 Japan
Phone: + 81 3 3496 7046
http://www.marumo-p.co.jp
Printing:Loyal Planning Co.,Ltd.

©Marumo Publishing Co.,Ltd.
Printed in Japan
ISBN 978-4-944091-47-8